Rationalizing Medical Work

Inside Technology
edited by Wiebe E. Bijker, W. Bernard Carlson, and Trevor Pinch

Marc Berg, *Rationalizing Medical Work: Decision-Support Techniques and Medical Practices*

Wiebe E. Bijker, *Of Bicycles, Bakelites, and Bulbs: Toward a Theory of Sociotechnical Change*

Wiebe E. Bijker and John Law, editors, *Shaping Technology/Building Society: Studies in Sociotechnical Change*

Stuart S. Blume, *Insight and Industry: On the Dynamics of Technological Change in Medicine*

Geoffrey C. Bowker, *Science on the Run: Information Management and Industrial Geophysics at Schlumberger, 1920–1940*

Louis L. Bucciarelli, *Designing Engineers*

H. M. Collins, *Artificial Experts: Social Knowledge and Intelligent Machines*

Paul N. Edwards, *The Closed World: Computers and the Politics of Discourse in Cold War America*

Eda Kranakis, *Constructing a Bridge: An Exploration of Engineering Culture, Design, and Research in Nineteenth-Century France and America*

Pamela E. Mack, *Viewing the Earth: The Social Construction of the Landsat Satellite System*

Donald MacKenzie, *Inventing Accuracy: A Historical Sociology of Nuclear Missile Guidance*

Donald MacKenzie, *Knowing Machines: Essays on Technical Change*

Rationalizing Medical Work
Decision-Support Techniques and Medical Practices

Marc Berg

The MIT Press
Cambridge, Massachusetts
London, England

© 1997 Massachusetts Institute of Technology

All rights reserved. No part of this book may be reproduced in any form by any electronic or mechanical means (including photocopying, recording, or information storage and retrieval) without permission in writing from the publisher.

Set in New Baskerville by The MIT Press.
Printed and bound in the United States of America.

Library of Congress Cataloging-in-Publication Data

Berg, Marc
  Rationalizing medical work : decision-support techniques and medical practices / Marc Berg.
      p. cm.—(Inside technology)
  Includes bibliographical references and index.
  ISBN 0-262-02417-9 (hc : alk. paper)
  1. Medicine—Decision making. I. Title. II. Series. [DNLM: 1. Patient Care Planning. 2. Decision Support Techniques. 3. Expert Systems. W 84.7 B493r 1997]
R723.5.B47   1997
610—dc20
DNLM/DLC
for Library of Congress   96-29283
                                                                        CIP

for Joy and Barthold Hengeveld
and in memory of Rebecca Berg-Kropveld

Contents

Preface ix
Introduction 1

1
The Withering Flower of Our Civilization: Reconceptualizing Postwar Medical Practice 11

2
Multiple Rationalities: The Different Voices of Decision-Support Techniques 39

3
Getting a Tool to Work: Disciplining a Practice to a Formalism 79

4
Of Nodes, Nurses, and Negotiations: The Localization of a Tool 103

5
Supporting Decision-Support Techniques: Medical Work and Formal Tools 123

6
Producing Tools and Practices 155

Notes 179
References 205
Index 235

Preface

This book is about decision-support tools (such as protocols, expert systems, and clinical decision analysis) and medical practices. It is a study of how medical practices are transformed by these tools and vice versa; of what "rationalizing medical work" does and does not look like. As such, it is the outcome of the coalescing of a sociological interest in what medical personnel do and a fascination with how formal technologies develop and function in concrete work practices.

Although it would be gratifying to locate the drive for these fascinations within myself, it would also be gratuitous. This book developed out of my Ph.D. thesis, for which I consider myself lucky to have had as supervisors the two persons who have most thoroughly shaped and sharpened my theoretical views: Gerard de Vries and Annemarie Mol. As a guardian of intellectual quality and consistency, Gerard is the thesis supervisor *pur sang*. I owe most of the hard times I had in the creation of this book to him, but I would not want to have missed his critical insights, sharp analyses, and didactical floggings. In her own unique way, Annemarie has been indispensable: her in-depth, critical, wide-ranging, yet always wholly supportive mode of supervising is everything one can hope for. I consider myself lucky, again, that we have gradually become colleagues and friends rather than supervisor and supervised.

Geert Blijham, my third supervisor, was vital in my getting access to the field of oncology, both practically and intellectually. He read and commented on several drafts of this book, and helped to ensure that I got the message down in clear terms. The way he could summarize the book in two sentences was always an important counterbalance to my tendency to take my work too seriously.

Harry Schouten has also been vital to my meanderings through oncology. I owe a great deal to our many discussions, the corrections he made whenever I got the "medical stuff" wrong, and his cheerful enduring of my nosy presence. The doctors, nurses, and patients I got to know during my fieldwork, and the tool builders and physicians I interviewed: without them, there would not have been a book.

Of the many people who have read, discussed, commented upon, or critiqued earlier portions and versions of the book, there are several that I especially want to mention. My (former) fellow Ph.D. students Antoinette de Bont, Monica Casper, Ruud Hendriks, Jessica Mesman, Susan Newman, Irma van der Ploeg, Stefan Timmermans, and Paul Wouters have been indispensable as both friends and critical readers. I have also benefited greatly from the support and helpful comments of Ed Berg, Wiebe Bijker, Geoffrey Bowker, Bruce Buchanan, Michel Callon, Adele Clarke, Diana Forsythe, John Law, Harry Marks, Randi Markussen, Bernike Pasveer, Leigh Star, and Lucy Suchman. Each in his or her own way has left a definite imprint on me and on the book. Saskia van der Lyke also left an important mark on the book—but the imprint she left on the rest of me is infinitely more valuable. In the last phases of this project, Hellen Heuts, Arjen Stoop, and Angel Waajen lent their helping hands, for which I am also grateful.

My daily workplace, the department of Health Ethics and Philosophy at the University of Maastricht, has developed into a congenial, supportive, and warm place to "come home to." The staff and students at the Centre de Sociologie d'Innovation in Paris have provided a highly stimulating—and challenging—environment in which to work on the final draft. Intellectually, I also owe much to the Maastricht-based BOTS, a research group on technology and society chaired by Wiebe Bijker. The Netherlands Graduate School of Science and Technology Studies (formerly LOOWTOK, now the Research School on Science, Technology and Modern Culture), albeit in constant flux, has provided me with high-quality and highly enjoyable workshops and seminars to sharpen my understanding of what the field of science and technology studies was about in the first place.

An earlier version of chapter 1 was published as "Turning a practice into a science: Reconceptualizing postwar medical practice," *Social Studies of Science* 25 (1995): 437–476. Fragments of chapter 5 appeared in "Modeling medical work: On some problems of expert systems in medicine," *ACM SIGBIO Newsletter* 14 (1994): 2–6; in "Working with protocols: A sociological view," *Netherlands Journal of Medicine* (in press); and in "Practices of reading and writing: The constitutive role of the patient record in medical work," *Sociology of Health and Illness* 18 (1996): 499–524.

Rationalizing Medical Work

Introduction

In the nurses' office on the oncology ward, Irene is looking for Mr. Field's medical record. Mr. Boottle, one of the patients she is taking care of, is suffering from the same type of throat cancer as Field. He will be treated with the same type of treatment—and Irene wants the protocol that describes it. She enters the medication room of the ward, where lists of antibiotic solutions are taped to the wall and little piles of scribbled chemotherapy recipes occupy the larger part of one of the desks. The record she is looking for is not there. Her colleague Matthew is there, however, and he accompanies Irene to where he last saw the record: in the residents' office.

The messiness of the residents' office is familiar. Patients' records and chemotherapy prescriptions are scattered about on the tables—there is even a prescription lying in the little basket that holds sugar and milk for coffee. Finding Mr. Field's record, Irene and Matthew routinely flip through it to where they know the protocol's scheme is summarized. Scanning the list, Matthew says: "You should change the Mannitol to Lasix, and do the Zofran intravenously on the first day. From then on, you only have to do it intravenously if necessary. And I guess you can just copy the timings they used here. That doesn't really matter much."

■

F. T. de Dombal on the making of a list of data items to be used as input for his computer-based decision-support tool (dealing with complaints of acute abdominal pain):

Some items we've never had money to investigate. That is why the x-ray of the abdomen is not part of our system: we've never had resources to find out whether it would make a difference. We've only had money to investigate the historical and physical examination data—and the system now only deals with those.

And when we had made the list, and the physicians really missed something they did not want to miss, we simply put it back in. So that's why [hospital X's] form has room for filling in the temperature, the pulse, and the respiration rate—although in 1920 somebody already said that these data were useless in acute appendicitis. But they want it, so we do it. They're only on the form, however: the computer doesn't use them. . . .

And when we began we ignored all kinds of mathematical niceties to get the thing going. There was not much around to hold on to. It was not a question of doing it by the book—we were making the book up as we went along. I wouldn't be trying to justify it. We couldn't go anywhere for help. There was Lee Lusted in the States—but nobody could get to the States in those days. Somebody told us about Lusted's book on Bayes. We read it, and we thought, well, yeah, that sounds reasonable. If somebody would have given us a different tip, we might have gone into a completely different direction. . . .

But don't get me wrong: we now have a system which has been running for years. In [hospital X] the system has achieved a false negative operation rate which is much lower than elsewhere. In addition, the system's efficiency is such that acute abdominal pain patients spend about half the number of days in the hospital here compared to elsewhere in the UK! (from an interview conducted on July 18, 1993; edited somewhat to enhance readability)

∎

In the last two decades, editorials and articles in medical journals have expressed worries about the current state of medical practice. At the start of an invited series of articles for the *Journal of the American Medical Association*, David Eddy argues that medical practice finds itself confronted with a profound challenge:

The plain fact is that many decisions made by physicians appear to be arbitrary—highly variable, with no obvious explanation. The very disturbing implication is that this arbitrariness represents, for at least some patients, suboptimal or even harmful care. (Eddy 1990b)

For example, the overall number of operations varies among geographical regions by factors ranging from 3 to 20. Physicians vary in what they do and how they reason: for example, when having to decide on the appropriateness of indications for coronary angiography or coronary artery bypass graft, UK physicians judge considerably differently than their US counterparts (Wennberg 1984; Brook et al. 1988).

According to these authors, the troubles are not surprising. The "ingredients needed for accurate decisions are simply missing for many medical practices" (Eddy 1990b). Often, there is no evidence available to determine the preferred action of choice. As a consequence, physicians "turn to their own experiences"—which are "notoriously misleading" (ibid.). Moreover, many authors argue that the problems modern doctors face are often too complex for ordinary humans to grasp comprehen-

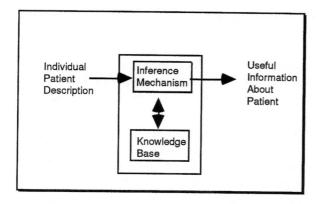

Figure 1
A simple model of a decision-support system. Source: Reggia and Tuhrim 1985. Copyright 1987 Springer-Verlag. Reproduced by permission.

sively. Medical decision making is hampered by "one of the most frustrating failings of the human mind . . . : its Lilliputian capacity for storing and retrieving important but infrequently used information" (Haynes et al. 1986, quoted in Wyatt 1991b). Eddy (1990b) agrees: ". . . even if good evidence were available, it is unrealistic, even unfair, to expect people to be able to sort through it all in their heads."

How to address this challenge? Many authors enthusiastically refer to so-called decision-support techniques as solutions: tools that can aid physicians in rationally practicing medicine. Different tools are mentioned: computer-based systems (such as the tool de Dombal was working on), protocols (or "practice policies"), and decision-analytic techniques ("clinical decision analysis").[1] These tools, which directly intervene in the medical decision making process, hold promise for an optimally scientific medical practice. Together with the medical practices they seek to rationalize—the workplaces of medical personnel like Matthew and Irene—they are the main characters of this book.

Schematically, a decision-support tool can be said to consist of two components, which may or may not be separated: a *knowledge base* and an *inference mechanism*. The former contains the knowledge needed to solve medical problems within a certain domain. Given a patient's description, the inference mechanism draws upon this knowledge base and generates advice tailored to this individual case (figure 1).

In *computer-based decision-support systems*, the inference mechanism can be embedded in symbolic "conditional rules" or in statistical formulas.

These tools are said to improve medical decisions by "efficiently placing current experts' knowledge at the doctors' disposal" and by helping physicians to "avoid errors caused by the imperfection of the human intellect" (Potthoff et al. 1988). Through such help, Schwartz (1970) argued, the "computer as an intellectual tool" can fundamentally alter the practice of medicine: it can "help free the physician to concentrate on the tasks that are uniquely human such as the application of bedside skills, the management of the emotional aspects of disease, and the exercise of good judgment in the nonquantifiable areas of clinical care."

Protocols, a second type of decision-support tools, offer pre-defined, stepwise, optimal paths through complex or troublesome medical situations. At each step, decision criteria can be built in to determine whether the next step can be taken: a protocol can be seen as a chain of generally simple conditional rules. Well-defined protocols should be able to "reduce inappropriate variation in services, improve the quality of care, and produce better health outcomes" (Field and Lohr 1990). By analyzing "decisions before the fact," protocols prevent the impossibility of having to rationally decide every time again from scratch (Eddy 1990c). These tools, Margolis (1983) proclaims, help us to "define, not for others, but for ourselves, the practice of rational medicine."

Finally, *clinical decision analysis* brings statistical techniques to bear on problems concerning the management of individual patients. The technique should allow the physician to choose the management strategy with the highest expected utility (see further). According to its advocates, it is the most "flexible, practical and yet rigorously logical approach" of all decision-making aids (Weinstein et al. 1980). "It is a logical, explicit, reproducible and objective process" that enables comparing alternatives and dealing with uncertainties; you can "examine in exquisite detail the assumptions and data on which the recommendations are based" (Detsky 1987). Its advocates address it as a veritable "new basic science in medicine" (Politser 1981) which is necessitated by the proliferation of medical tests and technologies and by the troubles physicians have interpreting statistical properties of data.[2]

First and foremost, decision-support tools would thus help to "transform the 'art' of medical decision making into a 'science'" (Komaroff 1982). They would make medical practice more rational, more uniform, and more efficient. Secondly, and closely related, the tools would help relieve the physicians' ever-increasing intellectual load. They would support the physician where he or she is weakest. Finally, the tools would achieve all this in a relatively unobtrusive and transparent way, simply by doing better what the physician had been doing all along. Expert systems, their advocates argue, reason as physicians do—they truly "understand" medical practice. Likewise, clinical decision analysis just makes the uncertainty and the risks and benefits inherent in each and every decision explicit. Decision-support techniques, it is said, fit medical work because they are structured like medical work.[3] The main difference is that the tool's "predetermined, formal, and explicit scheme of logic" (Komaroff 1982) allows us to *improve* upon this work. The tools, in other words, "can help us to demystify the practice of medicine, and to demonstrate that much of what we call the "art" of medicine is really a scientific process, a science which is waiting to be articulated" (ibid.).

Not everyone is enthusiastic about these tools, however. Several authors challenge both their feasibility and their desirability, contesting many of the advocates' claims. Hubert Dreyfus's line of argumentation in his influential 1972 book *What Computers Can't Do* (republished in 1992 as *What Computers Still Can't Do*) is typical of much of the criticism.[4] In his "critique of artificial reason," Dreyfus criticized computer scientists' attempts to achieve 'intelligent' systems—but the criticism is readily extended to all decision tools "using prespecified rules and principles" (Dreyfus and Dreyfus 1986). In the phrase that best captures the critics' central tenet, Dreyfus and Dreyfus state that such tools all "suffer from the impossibility of replacing involved knowing how with detached knowing that" (ibid.).

In most areas of human life, Dreyfus argues, rule-based systems cannot achieve an adequate level of expertise.[5] They can function in *formal* or *formalizable* domains separated from the messy real world, including tic-tac-toe, mathematical theories, and highly delimited areas such as the evaluation of a few clear-cut laboratory data. In these domains, behavior may be represented as strictly rule-governed. All relevant input and output is fixed; the mechanical processing of a set of rules can deal with all possible events. Outside these domains, the competence of rule-based systems is limited *in principle*. Here, rules or principles (if existing at all) always contain a *ceteris paribus* condition. "They apply 'everything else being equal', and what 'everything else' and 'equal' means in any specific situation can never be fully spelt out without a regress" (Dreyfus 1992).

To illustrate this argument, imagine a system that would classify an emergency department patient with acute abdominal pain into one of seven possible diagnoses according to a pre-set list of rules. The system (including the selection of the diagnostic categories) would be based on experiences with several hundreds of such patients.[6] It would then not be unreasonable to state that, in general, "everything else being equal," an emergency department patient with acute, sharp pain in his belly would have one of these seven diagnoses. Likewise, "everything else being equal," this patient's symptoms would lead the tool to classify the patient adequately. On the other hand, everything else might *not* be equal whenever events would occur that were not foreseen when the tool's rules or principles were encoded. If the patient were a pregnant woman, or if the patient looked psychotic, or if a neighborhood's water supply had been poisoned, the system's inferences would probably be hopelessly off. Trying to program all these possibilities into the tool is impossible, Dreyfus argues: the number of rules stating all the exceptions will soon explode. Moreover, how can we ever begin to try to tell the computer just when the basic rules apply, and when they might not? A computer program could not even begin to fulfill these tasks, according to Dreyfus: the domain of acute abdominal pain is just too large. Humans usually do not behave according to any clear-cut rules. As Wittgenstein (1958, sections 185–242) has argued, we often cannot even begin to make explicit *why* we say something is "equal" or not: we shrug our shoulders and simply say "we do that this way." In the end, the *ceteris paribus* condition points to the background of activities which is prior to all rule-like activity. This "life-form," which makes rule-like activity possible in the first place, can never be made completely explicit.

The conclusion of this argument, shared by many authors, is that human expert performance cannot be achieved with formal systems. Such tools grope for something that is fundamentally impossible. "The everyday world, it appears, cannot be described by . . . any finite number of any kind of statements," Blois (1980) states in the *New England Journal of Medicine*. Criticizing computer-based decision tools, Blois argues that the selection of facts relevant to a given situation is the physician's unique competence. Similarly, critics argue that the complexity of "the real context of clinical decision making" (Ingelfinger 1975) is such that decision-analytic techniques will have a hard time achieving "clinical relevance" (Cummins 1990). A proper analysis of a decision problem requires *all* possible outcomes, and *all* possible concerns and valuations about these outcomes, to be explicated in advance. Bursztajn et al. (1981), however, state that for both doctor and patient it is "difficult if not impossible . . . to include in the decision analysis all the various and sundry considerations that ultimately guided their actions, some of which they were not conscious of, some of which they would not have been comfortable dealing with explicitly and publicly, and some of which they never could have predicted would come up." Decision-analytic techniques, they state, are but poor representations of the complexities that go into real-time decision making. One cannot separate the decision from its context:

> When decision analysis separates facts from values, one cause from others operating along with it, an individual's welfare from that of the family or community, an illness from a person's life history, one decision from the context of other decisions that define its significance, it turns decision-making into something rather artificial and academic. (ibid.; see also Dreyfus and Dreyfus 1986)

Protocols, finally, have been under similar attack. Such rigid, predetermined schemes are said to threaten the physician's "art" by dehumanizing the practice of medicine and by reducing the physician to a "mindless cook" (Cutler 1979). By determining the path of action, the protocol renders the physician's skills superfluous. Physicians then merely have to be able to understand directions and to be willing to do what they are told: they are certainly not expected to think for themselves (Ingelfinger 1973). Moreover, such tools open the way for increased and uninformed control by "outsiders":

> There will be those in academia, in public health, in the insurance fields, in health maintenance organizations, and most surely and most terribly in the fields of law and government who will desire and will move to require strict conformity of practice to the presumed ideal. (May 1985)

Such coercion to a perceived "best way," May continues, stifles the life of medicine: it can become a form of "tyrannical domination" against which individual choices of physicians should be protected.

All in all, these critics argue, the tools' impoverished, codified versions of physicians' know-how do not do justice to the intricate, highly skillful nature of medical work. The idea of creating formal tools that make medical decisions is utterly mistaken. Every attempt to take the practical control of the decision process out of the physician's hands is doomed to fail—and is dangerous. Discussing medical expert systems, Lipscombe (1989) summarizes this issue as follows:

> ... the systems lack the flexibility of human interpretative processes, [thus] their use requires situationally specific human support. The current "control" design, however, provides no room for such support.[7]

■

The arguments of the advocates seem to be diametrically opposed to those of the critics: whereas the former stress the universal reach of the tool and the smooth fit between tool and practice, the latter stress the

8   Introduction

tool's strangeness and impossibility. In this book I take a different stance: I take the positions raised by advocates and critics as points of investigation, not as *a priori* assertions. Drawing on ideas and methodologies from science and technology studies,[8] I empirically study what "rationalizing medical practices" implies. The book is about the work—in a broad sense—needed to obtain and maintain a place for decision-support tools in medical practices.

To underscore the multidimensionality of these processes, the book switches among three interrelated levels of analysis.

First, focusing on discourses in *the medical literature,* chapters 1 and 2 investigate how views of medical practice changed so that the tools came to be seen as the solutions to the perceived problems. A central question, here, is how the conceptualizations of medical practice are linked to the tools that are supposed to mend its deficiencies. Some of the roots of the current ways in which medical work is addressed in medical journals and textbooks are traced, and the question is posed whether we are indeed witnessing the transformation of an "art" into a "science" or whether the image is more complex. What *is* this "science" we are heading toward? What is the "rational medical practice" that various decision-support tools are supposed to bring forth? Do the various decision-support tools embody a single, unitary notion of a rational practice?

Second, chapters 3 and 4 focus on the work of *system builders* to construct and implement some specific decision-support techniques. Drawing mainly on interviews and on participant observation, these chapters center on the negotiations these construction processes entail, and on how individual tools and practices are transformed in ways beyond anybody's or anything's grasp. Here, both advocates' and critics' positions are challenged: tools can be made to work, I will argue, but not because they fit so smoothly into the work of medical personnel. What does it mean, then, to get a tool to work in concrete practices? How do both practices and tools change? How does this affect the idea of medical practice's becoming "more rational"?

Third, chapter 5 focuses on *medical personnel* at work with the tools. Following them in their daily environments, this chapter looks at how the tools figure in and transform their work. What is their role after a decision-support technique has been installed? Do they become "mindless cooks," as many critics fear? Or are their tasks smoothly "supported," as advocates argue? And what does this mean with regard to who or what is "finally in control" of the decisions made?

In discussing the builders' struggles to get their decision tools to function, and the problems and challenges that medical personnel encounter in their work with the tools, I emphasize the inevitable divergences between the tool builders' claimed goals and their end products. Thus, I touch on issues of design, the standardization of work practices, "universality," and localization; on what medical work consists of; and on what happens when such practices meet the new tools. I address the question of what exactly we refer to when we speak of a "rational" practice and of "processes of rationalization," and the question of whether—even within the realm of formal tools—we should not speak of possibly diverging *rationalities*. I point out how these tools are never what they seem—without drawing the conclusion that they are useless and should be discarded. Throughout the book I draw attention to what these tools are and what they bring about, and to how medical practices are transformed by the tools and vice versa. In this way, some insight may be acquired into what is gained and what is lost when decision-support techniques are put to work.

The book does not contain the grand conclusions that may be found elsewhere. For example, decision-support techniques are not depicted as necessarily deskilling, or as always relieving medical personnel's task loads. One should be skeptical of utopian or dystopian accounts (Kling 1991): the worlds we are dealing with and the changes that we will be witnessing are far too complex to be reduced to such simplistic evaluations. This does not mean that this book is not evaluative in any way. On the contrary: it addresses the assumptions of both advocates and critics, and it takes a stance in regard to them. It also touches on their normative claims—including recurrent ideas on how decision-support tools should be evaluated.

Several techniques are unraveled and explored in the course of this book. The differences and similarities between them are manifold; however, each of them is a *formal tool* that uses a process of mechanized inference (a formula or a set of rules) to convert input to output.[9] In this way, they attempt to yield specific advice on the diagnosis and/or treatment of individual patients. Much more can and will be said about these carriers of what Abbott (1988) calls "commodified expertise." This description, however, serves as a way to delineate a class of tools from the much larger group of (social) technologies that can be said to pursue the "rationalization of medical practice"—technologies ranging from educational initiatives to the creation of electronic medical records and electronic textbooks. The latter developments, however influential and important, fall outside the scope of this study.

1
The Withering Flower of Our Civilization: Reconceptualizing Postwar Medical Practice

An 1982 editorial by Anthony Komaroff in the *American Journal of Public Health* opens with the following lines:

> Over the past 30 years, there have been increasing attempts to transform the "art" of medical decision making into a "science," to supplement a spontaneous, informal, and implicit set of judgments with the conclusions of a predetermined, formal, and explicit scheme of logic. The driving force behind this effort has been the perception that clinicians make medical decisions in an idiosyncratic manner, sometimes compromising the quality of care or wasting medical resources.

Komaroff speaks of a clear-cut project driven by an equally clear-cut need: a gradual and unswerving process of turning an art into a science, propelled by an awareness of the poor decision making capabilities of physicians. This rationalization project is achieved through implementing "explicit schemes of logic": decision-support techniques.

This depiction, recurring in many other articles and editorials, misses several crucial points. It takes for granted that the problem has simply been around for all those years, waiting for a solution that slowly crystallized in the form of the current decision-support techniques. It takes for granted, likewise, that postwar medical practice was an "art" waiting to be demystified; that it was "really a scientific process" waiting to be explicated (ibid.).

Focusing on the medical profession's discussion of these issues in postwar medical editorials and textbooks, this chapter presents a different view. It demonstrates that there was no single, unilinear process in which a previously "unscientific" practice became "scientific." Likewise, there was no single, stable realization of "a problem" that for thirty years guided the search for a solution. The "medical practices," "problems," and "solutions" discussed by editorials and textbooks were conceptualized differently within different discourses. What "scientific" medical practice is according to medical authors, and, concurrently, what medical practice's problems are, took on different forms in different times and contexts.

In this chapter, focusing on the United States in the period 1945–1990, I draw mainly on editorials in two leading American general medical journals: the *New England Journal of Medicine* (*NEJM*) and the *Journal of the American Medical Association* (*JAMA*).[1] In addition, I trace the introductions to some medical textbooks through different editions. To get a fuller grasp of the emergence of those discourses in which decision techniques became prominent, I explore their backgrounds more extensively.

Using sources such as editorials and textbooks obviously prohibits making claims about the way medical practices did actually change, or were actually structured, or how the discussions described affected the day-to-day work of practitioners. Editorials serve distinctive, rhetorical purposes in the creation and maintenance of "the profession"[2] and of the medical societies which the journals represent. In addition, my focus on the postwar literature leads to a relative neglect of earlier attempts to redefine medical practice as a scientific activity.[3] While these considerations are of utmost importance for a more explicitly *historical* approach to the topics discussed, they do not hinder the project set out here: grasping the transformations in postwar conceptualizations of "medical practice" in the medical literature.

As used here, the term "discourse" indicates an internally coherent way of depicting and describing medical practice, including its problems, the solutions to these problems, and whether (and in what sense) the practice of medicine is a scientific activity. Although the discourses have their origins and their moments of widespread acceptance in different periods, they did not simply replace one another. The various discourses sometimes argued, sometimes blended, and sometimes coexisted.[4] Also, the sequential discussion should not be understood as reflecting a gradual "enlightenment": the discourses were not steps toward a single view but disparate configurations. Moreover, each of these discourses can be seen to have operated in a circumscribed social and political time and space, indicating the historically contingent nature of their emergence and fluctuation.

I first characterize the discourse on medical practice and its problems that dominated the journals during the early postwar years. Subsequently, I zoom in on some of the divergent conceptualizations of "scientific" medical practice encountered in the material studied and its concurrent, divergent shortcomings.

Early Postwar Medical Practice: Its Nature and Problems

There are men and classes of men that stand above the common herd:
the soldier, the sailor, and the shepherd not unfrequently;
the artist rarely; rarelier still, the clergyman;
the physician almost as a rule.
He is the flower of our civilization,
and when that stage of man is done with,
only to be marveled at in history,
he will be thought to have shared but little
in the defects of the period
and to have most notably exhibited the virtues of the race.
—*Robert Louis Stevenson, quoted in Koontz 1959 and Talbott 1961*

The Science and the Art of Medicine

Postwar medicine in the United States witnessed a dramatic expansion of medical research, a growing number of physicians, and an increasingly powerful professional organization. It was a time of strong, optimistic belief in science as the American Way, and of massive increases in government funding for medical research, education, and hospital-building programs (Starr 1982; Stevens 1989). The early postwar editorials of medical journals express pride in the profession's accomplishments in providing medical care. Emerging "from a confusion of superstition and ignorance in its earliest days" (Anon. 1962), these editorials emphasize, medical practice has found a firm footing in the sciences of medicine. The achievements of these sciences in the first half of the twentieth century have been enormous:

> Never in the recorded medical history of the world have there been so many inspiring discoveries the importance of which has startled at times entire nations. ... Some diseases have almost been eradicated; at least they do not constitute serious health hazards. Other diseases are being brought under control at such frequent intervals that many persons have not grasped the enormity of such medical efforts. (Anon. 1950b)

Nevertheless, editors acknowledge that much remains to be done. We may have escaped the age of traditionalism (Anon 1956a), but our present age is "still stoutly laced with superstition and ignorance" (Anon. 1962). Inspired by the military and medical success of the coordinated research effort during World War II,[5] editors state that "all-over coordinated scientific investigation would accelerate the finding of needed truths that seem just beyond the horizon" (Anon. 1948b). With more high-quality research, the hope of seeing "alabaster cities gleam, undimmed by human tears" (Anon. 1956c) might become true.

Notwithstanding this jubilant confidence in the benefits that science would bring, medical practice is generally not regarded as a science. "The medical profession has three responsibilities," a 1952 editorial in the *New England Journal of Medicine* argues: "to cure the sick, to prevent disease and to advance knowledge" (Anon. 1952b). Of these three responsibilities, the editor addresses only the advancement of knowledge as a scientific activity. This separation between medical practice and science is pervasive in the early postwar years. "Science is nothing more than a method of reasoning equally applicable to the laboratory or the clinic," a distinguished physician states (Rutstein 1962). In mentioning the "clinic," however, he is not talking about medical practice but discussing the need for clinical research. Medical care is only aided by science; it is not a science itself. The practice of medicine consists of applying scientific medical knowledge to individual patients with unique symptoms and complaints. This application is an art requiring "medical ingenuity, experience, skill and individual attention" (Neal 1951). The relationship of the individual patient with his or her individual physician, inspired with the "spirit of dedication" (Rutstein 1962) is its core. Investigators have found that physical exertion is better for the cardiac patient than "the time-honored regimen of rest" (Goldwater 1959). But how do they apply this new scientific fact? "How do physicians judge whether or not a cardiac patient can safely undertake a particular job? The answer is the same as it is to many questions in medicine. . . . Clinical judgment, common sense, and courage comprise the key to management." (ibid.)

In the 1950s, when editors proudly spoke of "scientific medicine," they were proud of the *availability* of "scientifically" generated medical knowledge and technology. The phrase "scientific medical practice" was seldom used, and when it was it rarely meant anything other than the presence of scientific knowledge. Moreover, when medical practice was addressed in terms of its scientific character, it was often in a critical tone. Medical care has paradoxically been jeopardized by the same advances of medical science that have helped it, authors state. The increase of "scientific paraphernalia" threatens to turn physicians into mindless technicians. The overabundance of recently discovered laboratory tests or "so-called miracle drugs" (Neal 1951) can lead to a "'pushbutton' type of decerebrate medicine" (Alvarez 1953). An editor makes this point poetically:

> . . . the mechanization of an art, a skill or a culture cannot adequately take the place of the personal sensitivity of its followers, which had made a living thing of it, nor can the interpretation of a physical phenomenon replace the sympathetic understanding of a total human problem. The air brush is not a complete substi-

tute for a bundle of camel's hair, and the drawing of a horizon with a rule cannot convey exactly what the artist sees. (Anon. 1952a)

The relationship between the science and the practice of postwar medicine was fundamentally ambiguous. On the one hand, editors argue that "a generation or two ago the physician had no choice but to rely largely on his skillful exploitation of the art of medicine" (Talbott 1961). The art of medicine, to be successful, requires a science to be applied. On the other hand, medical practice should never let this science impinge too close—lest the ability to artfully apply this science might be lost. Nothing is more important than "the authentic 'feel' of the case—the appreciation of the degree of the patient's pallor, the odor of her breath, the precise color of the blood-tinged sputum, the dryness of the parched tongue, the querulous reaction to her illness and the psychosomatic background for this reaction" (Anon. 1951a).[6]

The Problems of Medical Practice

Early postwar editorials did distinguish many problems. The costs of medical care were rising, and more and more people had trouble paying for their medical bills. Also, the number of physicians was felt to be inadequate and unevenly spread across the nation. One editor noted in 1948 that "from many sections of the United States complaints have come lately that persons who have called physicians late at night have been unable to secure attendance. . . . In one western community a fire chief gave to the press information to the effect that he had called twenty-four doctors and had been unable to secure attendance by any of them." (Anon. 1948d)

And when physicians are available they often make "errors in judgment and errors caused by ignorance or misinterpretation of the patient's condition"—leading, for example, to the performance of "unnecessary operations" (Anon. 1948a). While not willing to underwrite the criticism from "outside" the profession, another editor cautiously confirms that "medical service is far from perfect." "Is there a shortage of doctors? Is 'socialized medicine' the solution for our medical problems? Is there too much specialization? Why is it hard to get a doctor in an emergency? How can the cost of medical care be reduced?" (Anon. 1948c)

These problems, it is often argued, share a common root. "In the practice of medicine . . . , technic has outrun the present capacity of society to place scientific progress at the service of mankind. At this point while the spirit may be considered as willing, the social and economic flesh presents its weaknesses" (ibid.). These external restraints, editors argue, become more and more salient now that medicine's scientific achievements

16   Chapter 1

continue at unprecedented speed. "The success of the medical profession in increasing the life span has resulted in a whole new set of socioeconomic problems," and older patients require "more detailed care at greater cost to the hospital" (Oughterson 1955).

The "errors" and "unnecessary operations" are also commonly explained in this vein. The conditions under which general practitioners in poor urban areas have to work "permit only a superficial and unsatisfactory approach to the problems of diagnosis and therapy" (Collings and Clark 1953). In addition, "ineffective or obsolete methods of practice established by tradition," like "the fixed idea that some form of treatment must be given," can persist, since they are "forced on the profession by the majority of its patients" (Anon. 1950c).[7] It is the socio-economic context of medical practice that is to blame.

One of the external obstacles to optimal medical practice is the perceived increase in governmental involvement and regulation. In the eyes of the medical profession, "political control or domination of medical practice" (Anon. 1945) is an evil to be avoided at all costs:

When the Government finances and operates medical services, three things that are essential to good care are inevitably compromised: First of all, the wholly voluntary relation between the doctor and the patient... [Second,] complete privacy, with no intermediary between the patient and the physician. [Third,] maximum incentives for the doctor to do his best work and constantly to improve himself. (Judd 1960)

Against the backdrop of Communist Russia, the atrocities of Nazi science, and the general demise of trust in government power, the image of the free, autonomous professional was a powerful one (Starr 1982; Rothman 1991). (See figure 1.1.) Increasing governmental control "would seriously handicap, if not abolish, many existing enterprises that are gradually accomplishing the very things that contribute to the improvement of medical care" (Anon. 1946). It will inevitably lead to "the assembly line type of practice so greatly deplored elsewhere" (Anon. 1952c).[8]

The causes of medical practice's problems were thus peripheral to the physician's art. Moreover, editors argued, the problems themselves were marginal. "The general health of the population of the United States is constantly improving," an editorial argues in 1950. "No one can deny this without resorting to falsification. Those who claim that a health crisis exists in this country cannot prove it." (Anon. 1950a) Another editorial (Anon. 1961b) insists that "the great bulk of medical care is handled efficiently and economically." "To the credit of the medical profession," practices where this is not the case are "exceptional" (Anon. 1947b). Often, it is argued, the so-called problems are in fact merely patients'

There can be but one master in the house of medicine, and that is the physician.

Figure 1.1
A cartoon showing the proper role of the physician according to an editorial. Source: Anon. 1938. Copyright 1938 American Medical Association. Reproduced by permission.

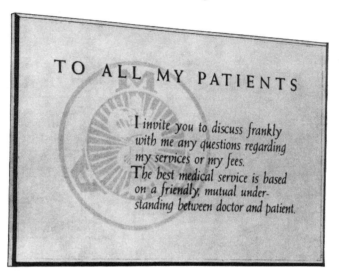

Figure 1.2
A solution to medical practice's problems in the postwar years. Source: *Journal of the American Medical Association* 147 (1951), no. 11, adv. p. 47. Copyright 1951 American Medical Association. Reproduced by permission.

misconceptions. "The vast majority of patients' grievances against their doctors stem from one thing—misunderstanding. . . . Patients must be encouraged to talk over with their doctor any questions they might have regarding his services or his fees."[9] (Anon. 1951b; see figure 1.2)

In this early postwar discourse, then, medical practice was seen as the artful application of medical science. To address its problems, editorials did not call for a more "scientific" medical practice, as they would some decades later. At most, authors stressed the necessity of a "total view" (Doyle 1952) of "diligence and sound judgment" (ibid.) and of a "careful and judicious" evaluation of the patient (Goldwater et al. 1952). Solving "the present-day inadequacies of medical care" required addressing "various socioeconomic factors with which the majority of physicians have had little experience and over which they have little control" (Anon. 1946). The origins of the problems of medical practice were seen to lie in the existence of restraints external to the practice of medicine itself: the fight was against corruption seeping in at the margins. Once these restraints were removed, once governmental control was diminished and more high quality physicians were trained, the practitioners' art would thrive undisturbed.

Guidelines for the evaluation of disability (Anon. 1958) were the one decision-support-like instrument referred to in editorials in these years. These guidelines would help physicians find their way through the labyrinth of complex demands made by various institutions. Doctors "are often confronted with a maze of zigs and zags formed by the differing and fragmentary concepts of diverse agencies, such as the Veterans Administration, Social Security Administration. . . . Now, at last, the zigs and zags are being pieced together into straighter lines." (ibid.) In agreement with the discourse described here, these guidelines are intended to overcome troubles generated by agencies *external* to medical practice. It is again the weakness of the socioeconomic flesh which is at stake: the physician's position as the flower of our civilization remains untouched. An editorial (Anon. 1959) proudly quotes President Eisenhower's tribute to the medical profession: "We shall one day establish a world community of peace-loving nations," the president says in an address to the American Medical Association. "In bringing about this happy result," the president continues, "no one can or will do more than the doctors of medicine."

The Multiple Sciences of Medical Practice

In the postwar years, however, other typifications of medical practice started to emerge. Editorials from the 1950s and 1960s resonated with an awareness that medicine was living through its "Golden Years": as mentioned above, research efforts were exploding in size and number, and the medical profession's ability to frame debates on the future of health care was stronger than ever. The intertwining of professional power and scientific development seemed to hold the promise of indefinite accomplishment and strength.

Against this background, "scientific" became a positive (albeit as yet rather vague) denotation for medical practice.[10] Science "permeates" medical practice, it was argued—the boundaries between the sciences that generate and the physicians who apply medical knowledge should not be drawn so rigidly (Anon. 1947a). More and more, authors and editors spoke about "scientific medical practice" in a way that added connotations that went beyond the mere application of scientific knowledge. What these connotations would be, however, was not evident. "Scientific" is not the unequivocal attribute it would seem to be. What "scientific medical practice" actually meant could be and was answered in more than one way; what specific problems would come into focus differed concurrently. In this section, I describe some subsequent discourses on medical practice and its problems. The different depictions of the "scientific" nature of medical practice, it is argued, throw their distinctive shadows—and contain their distinctive solutions.

Situating Science as the Foundation of Practice
First, a discourse can be discerned in which "scientific medicine" denotes a configuration that is essentially as described above: a foundation of scientific medical knowledge, authored by the basic (and, increasingly, the clinical) sciences, skillfully applied in unique patients. Here, however, some problems that had been absent from the discourse discussed above are centralized. Elaborating themes dating back to the turn of the century,[11] editors point out a deficiency *within* medical practice as the cause of these problems. Attention was drawn to recurrent, specific problems of communication between physicians and researchers.

For one thing, editorials stress that, as a result of the rise of clinical and epidemiological research, clinical observations have become "irreplaceable scientific data." Therefore, "scientific precision" (Anon. 1965d) is crucial in their gathering and classification. But, it is argued, medical practice lacks such precision: it is characterized by terminological chaos and the "loosely catalogued, cross-referenced and integrated" state of clinical records (Anon. 1950d). Because of "prejudiced, incomplete, or noncomparable reports," clinical research often fails "to answer a single question" (Howard 1961), thus hampering "the progress of medicine" (Anon. 1950b). In international studies the situation is even worse. A comparison in 1960 between the "cardiopulmonary semantics" of Britain and United States reveals "striking . . . discrepancies in terminology" (Meneely et al. 1960). For example, the common term "chronic bronchitis" refers to a benign, low grade inflammatory process in the United States, while it indicates a potentially grave and progressive pulmonary disease for "the British." How can international research efforts come to anything when this "Tower of Babel situation" (ibid.) persists?[12]

As a possible solution to these problems, standards are created that set the minimum requirements of medical-records systems (Anon. 1961b).

In addition, the new, "scientifically designed and arranged" *Standard Nomenclature of Disease* is praised. It describes "the main characteristic about every disease" with a "clarity that far surpasses words and with unambiguity by means of a specific code number for each disease—a code number in which each digit conveys a factor without confusion." With this tool, "the comparison of clinical experience between institutions becomes sound, is easily encouraged, and is attainable as uniformity and accuracy of clinical expression are prescriptive and customary" (Anon. 1957b).

Besides researchers, "health care providers, planners, government agencies, and epidemiologists" (Froom 1975) will also benefit from these developments: all these groups depend on "meaningful statistics" (Anon. 1954b). For all these groups, in addition, it is argued that "precise definitions are particularly important now that computers offer the promise of significant assistance in data retrieval" (Anon. 1967).[13]

Finally, medical practice will benefit. Standardization of terminology ensures that "hospital records are easily understood by all who read them, and physicians can understand what others are talking about when diseases and operations are mentioned" (Anon. 1954b). Also, improved communications will help to fight diagnostic errors and observer variation (see, e.g., Garland 1959). Many treatments are applied without "proper data" to support the effectiveness of the therapy, an editorial argues. The "procedures of diagnosis and treatment performed by the physician should have at least as much validity as is required of a drug company before a new drug may be offered on the market" (Anon. 1963b). Standardizing terminology and records will enhance the ability to apply the results generated by such research.[14]

What kind of "scientific medical practice" do we see here? The authors and editors quoted have not moved away much from the image of medical practice described in the previous section. For them, the scientific status of medical practice still derives primarily from the foundation of scientific knowledge it rests upon. What has been added is a set of new problems, due to strained communication between researchers and physicians and among physicians themselves. These troubles are seen to be caused by a weakness of medical practice itself: the practice of medicine lacks uniform records and terminology. Through standardizing nomenclatures and record-keeping procedures, medical practice can improve both the production of data needed for medical research and the application of the results of this research. Such measures will align individual efforts and increase efficient and precise communication; they will fit medical practice more smoothly into the rapidly expanding postwar research enterprise (Marks 1988). In this way, medical practice can become "scientific" (now not a negative term), providing the optimal, artful application of scientific knowledge to the needy patient.

Situating Science in the Structure of Medical Action

In the view of scientific medical practice just described, medical action plays a secondary role. It produces the data upon which much medical science feeds and applies the scientific knowledge thus generated.

A different rendering of medical practice emerged in the late 1960s and the 1970s with the often-quoted works of Alvan Feinstein and Lawrence Weed. Here, medical practice is not primarily the application of a science located elsewhere: the practice of medicine itself is depicted as a scientific activity. "Clinicians do not usually regard ordinary patient care as a type of experiment," Feinstein argues in his classic work *Clinical Judgment*. Nevertheless, "every aspect of clinical management can be designed, executed, and appraised with intellectual procedures identical to those used in any experimental situation." At the bedside, "exactly the same principles of scientific method" apply as to "any other experiment" (Feinstein 1967). Likewise, Weed (1968) states that "the practice of medicine is a research activity." For these authors, medical action *has the same structure* as scientific action:

> The scientist defines a problem clearly, separates multifarious problems into their individual components, and clarifies their relationships to each other. He records data in a communicative and standard form and ultimately accepts an audit from objective peers by seeking publication in a journal. (Weed 1971)

According to Feinstein and Weed, each and every step occurs in the clinical setting as well. Although the physician's actions in medical work may differ in practical shape, a doctor defines, separates, clarifies, records, and audits just as much as a scientist.[15]

Depicting medical work as a "scientific" activity in itself, these authors introduced a general, explicitly normative framework. With this framework, medical practice could now be scrutinized and judged. When, as above, "scientific medical practice" means the optimal usage of scientific knowledge, one can scrutinize medical practice for places where this knowledge is improperly used or not used at all. Standardization is then a way to guarantee optimal flow of information so that the benefits of science reach those who need it. By explicitly labeling the distinctive steps in the clinical process as elements of the scientific method, however, Weed, Feinstein, and others made medical practice analyzable in a new and thorough way. The individual steps of "the experiment," the definition of the starting point, the planning of the intervention, and the observation of the outcome could now be discerned *and* judged.

Redescribing medical work in this way, concurrently, yielded new problems and new solutions. The deficiencies were now not merely "problems of communication" but inadequacies affecting the heart of medical practice. Since medical action is often not recognized as *consisting of* clinical experiments, these authors argue, medical practice largely lacks "the scientific qualities of valid evidence, logical analyses, and demonstrable proofs" (Feinstein 1967). The explosion of diagnostic tests and therapeutic possibilities, combined with the increasing specialization and (thus) cooperation of physicians, has drastically changed the daily work of the clinician. Now, these authors say, medical activities concerning a specific patient's problem are often thoroughly spread out in time and space. At the same time, physicians are wielding more and more powerful therapeutics, with a greater capacity to do harm (ibid.; Hurst 1971). Feinstein, Weed, and others agree that the need for more standardized terminology and procedures is urgent. Without it, there is no way that medical practice will ever be able to properly adhere to the rules of the scientific method.

Several interrelated strategies were proposed to bridge the gap between the state of actual practice and its scientific potential. Weed and his followers argued that a basic step toward a solution lay in a new, standardized type of medical file. In the "problem-oriented record," all data, action plans and progress notes of all personnel involved are to be organized around the problem(s) of the patient. In this way, physicians can *act scientifically*. Through the problem-oriented record, the doctor "is able to organize the problems of each patient in a way that enables him to deal with them systematically" (Weed 1968; Hurst 1971).[16]

Others were attempting to automate the medical history: a computer would ask a long list of standardized questions of the patient (individualized, somewhat, through branchings in the list), and would subsequently present the answers to the physician in a summarized form (see figure 1.3; see Barnett 1968 and Mayne et al. 1968).

Weed's dream was to put all these intertwining efforts together with an automated version of the problem-oriented record, so that "all narrative data presently in the medical record can be structured, and in the future all narrative data may be entered through series of displays, guaranteeing a thoroughness, retrievability, efficiency and economy important to the scientific analysis of a type of datum that has hitherto been handled in a very unrigorous manner" (Weed 1968).[17]

```
┌─────────────────────────────────────────────────────────────────┐
│ Which of the following phrases best describe the speed of your heartbeat? │
│                                                                 │
│  ☐   I am not usually aware of the speed of my heartbeat.       │
│  ☐   My heartbeat is sometimes very fast.                       │
│  ☐   My heart seems to beat very fast all of the time.          │
│  ☐   My heartbeat is sometimes very slow.                       │
│  ☐   I occasionally have attacks of very rapid heartbeat,       │
│      which usually start suddenly and stop suddenly.            │
│  ☐   None of the above describe it.                             │
│                                                                 │
│  ☐               ☐             ☐                                │
│ ←Go back        Erase         Continue→                         │
└─────────────────────────────────────────────────────────────────┘
```

```
┌─────────────────────────────────────────────────────────────────┐
│ Yellow jaundice is a condition in which the skin and the whites of the eyes │
│ become distincly yellow. When jaundice is present the bowel movement │
│ may become a pale putty color and the urine may become dark in color. │
│ Do you understand?                                              │
│                                                                 │
│  ☐   Yes                                                        │
│  ☐   No                                                         │
│  ☐   I understand the explanation but I still don't understand the question │
│                                                                 │
│  ☐                                                              │
│ ←Go back                                                        │
└─────────────────────────────────────────────────────────────────┘
```

Figure 1.3
Screens showing two of the more than 200 questions of an automated medical history and some "supplemental explanations." Source: Mayne et al. 1968. Copyright 1968 *Mayo Clinic Proceedings*. Reproduced by permission.

For Feinstein, the basic problem was the unstandardized language physicians use, and the unstandardized procedures through which they acquire their data. For example, physicians will erroneously write "dyspnea" when they refer "to the fallacious 'shortness of breath' described by a patient whose only symptom is inspiratory chest pain."[18] It was crucial to oppose the usage of the "gaggle of eponyms, synonyms, and terms, which, although declared officially obsolete 30 years ago, still bob up with amazing persistence" (Anon. 1954a). Thorough uniformity in terminology, "objective preparation," "delineated precision," and "standardized interpretation" in the acquisition of clinical data are essential if medical practice's scientific potential is to be fulfilled (Feinstein 1967).[19]

Given these basics, algorithms (detailed, graphic protocols) could subsequently structure the diagnostic and therapeutic processes themselves. With the conceptual apparatus in place, the protocol could improve the sloppy design and execution of these scientific experiments called patient care. Feinstein (1974) explains their role eloquently:

> Until recently, a clinician who wanted to retain the traditional "art" of diagnostic reasoning could not avoid its concomitant scientific aphasia. Having no symbols, no structures, and no tactics with which to demonstrate his patterns of thought, he could not attempt to express his reasoning with any of the traditional oral, written, or graphic patterns of scientific communication. A chemist could use chemical formulas, drawings, and arrows to show the path of an enzymatic transformation; a physicist could use photographs to show the path of an electron's movement; but a clinician had no substance or method that could show the path of a rational sequence. . . . [With the recently developed techniques of] algorithms, flow-charts and decision tables . . . a clinician can now, at long last, specify the flow of logic in his reasoning. [With these], diagnostic reasoning can begin to achieve the reproducibility and standardization required for science.

In the influential work of Feinstein, Weed, and others we thus find another discourse on medical practice and its problems. "Scientific medical practice" here denotes more than the scientific base of that practice, the existence of scientific knowledge. Medical practice is itself declared a scientific activity, since its actions are structured like scientific action itself. The way the different steps of the scientific method are executed, however, leaves much to be desired. The increasing specialization, and the subsequent increase in actors dealing with one and the same patient, will only aggravate the problem.

As in the previous discourse, these authors argued for standardization measures. Here, however, both the goal and the scope of standardization have changed. Standardization is now seen as a fundamental prerequisite to scientific medical practice: it is a *sine qua non* for the full development of this new science. What is being standardized changes accordingly: creating standard record-keeping procedures and a uniform terminology is only the first step. Protocols can subsequently structure the ongoing process of medical work. With this depiction of the nature of medical practice, a niche is created for a tool supporting the core of the physician's task: the scientific work of diagnosis and treatment.[20]

A Turn to the Mind: New Sciences of Medical Practice
For the two discourses just described, the notion of a "scientific medical practice" was closely intertwined with the idea of standardization. And standardization efforts were not new to medical practice. In the first

decades of the twentieth century, Frederick W. Taylor's "scientific management" of work processes had been exported from the factory floor to a wide range of other domains—including medicine. The idea of increasing efficiency and the control of production by breaking down complex tasks into standardized subtasks had been translated into attempts to standardize medical writing, medical education, hospitals, and record-keeping procedures (Stevens 1989).[21] These early endeavors, however, disappeared in the decades that followed. With the increasing resentment against government and hospital administration involvement with physicians' work, the standardization practices of the American Medical Association and the American Hospital Association were largely limited to setting standards for residency training and for the organization of hospitals' medical staffs. They did not address medical procedures (ibid.).

The renewal of explicit attention to standardizing aspects of medical work coincided with the beginning of what Starr (1982) calls "the end of the mandate" of the medical profession. Calls for health-care reform again started to increase, and public outcries about skyrocketing costs, poorly distributed and poorly organized care, and the self-centeredness of the medical profession slowly began to displace the optimistic belief in medical progress.[22] Editorials in medical journals reflected some of these worries. "It has become fashionable in academic circles to raise doubts about the effectiveness of all medical care," an editor argued, but "this is an understandable and needed reaction against the therapeutic enthusiasm that has been rampant in the profession and the general population for the past 25 years" (Haggerty 1973; cf. Good 1995).

Yet the renewed focus on standardization evoked mixed feelings. Standardization, some authors felt, would also benefit hospital administrations, insurance companies, and government agencies, thus playing into the hands of those who were most threatening to the fervently defended professional sovereignty. The notion of such impending regulatory measures conflicted with the image of the autonomous, free professional. In 1977, when standard protocols for the collection of laboratory data were being offered to practicing physicians, an editor retorted: "It is a small step from 'offer' to 'provide.' It is another small step from "provide" to "require." And all of this would finally be enforced with the clout of Medicare. Be alert. The first sproutings of the bureaucratic liana may be no farther away than your own clinical laboratory." (Crosby 1977)[23]

While these polemics went on, other notions of what would constitute a scientific medical practice came to the fore. In the 1970s and the 1980s, new discourses became prominent in which the scientific character of medical practice became a thoroughly *individualized* notion. Rooted in the booming field of cognitive psychology, these discourses contained an image of medical practice that perfectly fitted the profession's vision of the autonomous physician. Here, medical practice's scientific character was not a feature of a social, collective practice. With the shift away from the alignment of individual efforts, the focus on standardization faded. In these new vocabularies, the scientific status of medical practice was redefined as a feature of the physician's mind.

The "cognitive revolution" in psychology had reestablished the human mind's active, constitutive role in human behavior—a role that had been out of the spotlight when behaviorism had been dominant.[24] In *Human Problem Solving*, one of the central works in this reappraisal, Alan Newell and Herbert Simon (1972) developed the argument that human minds can be seen as information-processing systems. Humans, these authors state, form symbolic mental representations of problems that are presented to them and then solve the problems by logically processing the symbols—that is, by moving and modifying chunks of information.

In 1978, Arthur Elstein and his co-workers brought this approach to the attention of the medical profession. Drawing heavily upon Newell and Simon's work, they called their influential book *Medical Problem Solving*.[25] Like Newell and Simon, Elstein et al. regard problem solving as the symbolic processing of information. Unlike Newell and Simon, however, they state from the outset that medical problem solving is a "hypothetico-deductive" process. The physician, Elstein et al. argue, mentally tests hypotheses. Generating possible hypotheses (i.e., diseases or disease categories) from a few symptoms and signs, the physician subsequently chooses from the hypotheses by testing them against other symptoms and signs obtained from the patient.

Interestingly, Elstein's team thus introduced a terminology strongly affiliated with notions of the process of scientific reasoning. Problems are solved by the logical processing of symbols, Newell and Simon had argued; "scientific reasoning" is not called upon in their work. However, in medical education, the field of Elstein and his co-workers, the "hypothetico-deductive method" was a familiar notion. Since the mid 1950s it had been seen as a useful technique for teaching students clinical reasoning (Groen and Patel 1985).[26] In the process of incorporating Newell and Simon's work, Elstein et al. turned what had been a teaching technique into an explicit and detailed *description* of medical problem solving.

In this way, *medical action was equated with scientific action, which, in turn, came to denote a specific thought process.* "Medical problem solving," wrote

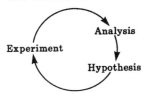

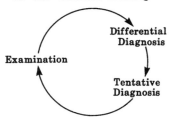

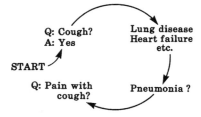

Figure 1.4
A 1984 medical textbook's vision of "the clear parallel between the general scientific method and the scientific method applied in the clinical setting." Source: McGehee Harvey et al. 1984. Copyright 1984 Appleton and Lange. Reproduced by permission.

Kassirer and Kopelman (1988), "is a progressive building process of hypothesis generation, testing and verification—devoid of intuitive, inspirational leaps."[27] It is not hard to see how the attractiveness of this conception lay in the match it created between the notion of the autonomous physician and the image of medical practice as a scientific activity (figure 1.4). The early postwar discourses' opposition between "practice" and "science" had dissolved completely—but both notions had changed in the process. Medical practice, here, is the reenactment of "the incremental process . . . [of] scientific discovery" (ibid.), and the scientific character of medical practice has become a mental category. "Training the mind" becomes crucial; "just as the researcher draws conclusions from experimental data, the practicing physician must base his decisions on the analysis of the medical history, physical findings, and laboratory tests" (Greenberg 1978).

The notion of the physician as a processor of symbols, however, is only one variant of the cognitivist discourses on medical practice. The other main variant also locates the scientific status of medical practice in the mind. Like the information-processing approach, it turns medical practice into a science while concurrently underwriting the profession's image of the individual, autonomous physician. Here, however, medical diagnosis is seen as an exercise in *statistics*.[28] Medical problem solving, various authors argue, necessarily involves calculation—whether explicitly or not. "Since 'chance' or 'probabilities' enter into 'medical knowledge', then chance, or probabilities, enter into the diagnosis itself," Ledley and Lusted (1959) argued in a seminal article in *Science*. They continued: "The reasoning foundations of medical diagnosis and treatment can be most precisely investigated and described in terms of certain mathematical techniques. [These] are inherent in any medical diagnostic procedure, even when the diagnostician utilizes them subconsciously, or on an 'intuitive' level."

Authors building upon these ideas often invoked notions from economics. In the 1970s and the 1980s, economists' terminology spread into the editorials of the leading American medical journals. Reflecting an increase in public attention to health care's costs and manageability, notions like "cost effectiveness" and "marginal utility" became more and more prevalent. Physicians were told to learn more of the "dismal science." Riesenberg (1989) noted that "patients deserve it, payers seek it, society needs it, and common sense dictates it." Schwartz et al. (1973) stated that "the essence of clinical judgment resides in the ability of the physician to weigh the advantages and disadvantages of a diagnostic or therapeutic procedure and to choose a course of action for a particular patient based on estimates of such costs and benefits."[29] Here the potential scientific status of medical action is found in its exact, quantitative, calculable nature.

The two approaches often fight for hegemony over the terrain of medical judgment. Authors within the statistical approach often accuse the symbolic-information-processing theorists of remaining too vague. Talk of "chunks of information" lacks quantitative precision: "Only statistics seem able to handle the high degree of variability and the complex pattern of correlated data. With so many variables, practically nothing comes clear and distinct; there are thousands of combinations possible. So calculating the odds of this and that, given such and such, seems the only way to have a reasonable basis for decisions." (Clouser 1985) In complex situations, the mind can reach a solution only by means of statistical techniques—whether it does so consciously or not. "The need for accurate probability assessments cannot . . . be avoided" (Borak and Veilleux 1982).

Many advocates of the information-processing approach deny that the physician is an "intuitive statistician" (Elstein et al. 1978) or that physicians "grow decision trees in their heads" (Moskowitz et al. 1988). These are ridiculous assumptions, they argue. In their view, statistical models should not be understood as a description of physicians' problem solving or decision making. All they do is "predict" or "black-box like simulate" this process (Elstein et al. 1978), which, at heart, consists of non-quantitative, symbolic reasoning. "Mathematical psychologists" are wrong to assume "that the kinds of mathematics that work in physics and chemistry will fit the requirements of psychology," Baars (1986) continues. "It is assumed that quantification is important for representing psychological phenomena. It is quite possible, however, to get mathematical precision without quantification." (ibid.)[30]

Within this cognitivist reconfiguration of the shape of medical practice, then, we see two competing discourses. One approach views the physician as mentally manipulating non-quantitative symbols according to complex processing rules. The physician follows the hypothetico-deductive method in solving a patient's problem. The other approach argues that physicians intuitively calculate their way to a decision, and that the scientific status of medical practice resides in its exact, mathematical character.

Although the differences between them are substantial, these discourses have much in common. With the approaches described here, the "scientific nature" of medical practice has become a *mental* category. What makes medical practice a science is not the fact that the shape of medical work follows positivist rules of scientific method, as in the discourse exemplified by Feinstein and Weed. Rather, medical practice is a science because the physician's individual thought processes (consciously or not) exemplify scientific reasoning. What medical work is, then, is again reshaped. Medical work is not seen as a social activity, but as an individual, cognitive process.

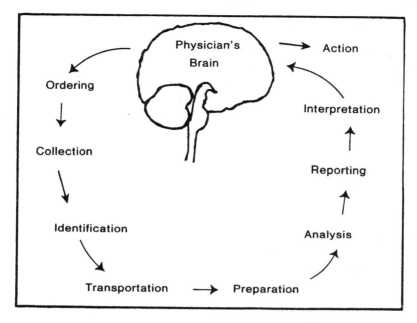

Figure 1.5
The physician's brain as the core of medical action. Source: Lundberg 1981. Copyright 1981 American Medical Association. Reproduced by permission.

We have come back to the notion of the individual, autonomous professional we saw in the early postwar discourse—but the role of this individual has changed radically. No longer is physicians' work depicted as the authentic art of a Man of Character. The individual's merit has shifted from "character" to "method": since the autonomous physician mentally follows the rules of science, (s)he has become the quintessence of the science called "medical practice."[31]

A Turn to the Mind: New Problems and Solutions

In the past few days, I have been feeling as if my calvarium and that of my fellow clinicians has been raised a little. A lot of people have been peering in to the interstices of the doctor's not always efficient cerebral cortex. This collaborative study of the interesting routes we take in making a clinical diagnosis promises to be fruitful, a trifle revealing and, I hope, not too embarrassing.
—E. D. Pellegrino (1964)

In the cognitivist discourses, medical practice is again described as a scientific activity—albeit in a new way, generating a new picture of the nature of medical practice. Again, the models used to describe medical problem solving function, at the same time, as a normative framework. New problems are highlighted and new deficiencies spotted. Actual physicians' behavior is compared to and judged by the model, and, subsequently, the model explains the flaws found (cf. Gigerenzer and Murray 1987). Seen from this angle, these new discourses on medical practice turn out to be something of a Trojan Horse. They open up medical practice to a new type of scrutiny, undermining the stamp of scientificness they had just established. The physician's mind, the new locus of the scientific character of medical practice, appears to be fundamentally deficient, whether seen from the statistical or the information-processing perspective.

It is noted, for example, that physicians are sloppy in their adherence to the scientific process. It is very difficult for a physician to "avoid wishful thinking or introduction of fantasy," since "we are inclined to form opinion before we have all the facts" (Danilevicius 1975). "Typically, physicians form tentative hypotheses (diagnoses) from a small amount of data. As new information is acquired, they sometimes ignore new data that contradict hypotheses currently remembered" (Politser 1981).[32]

Authors within the statistical discourse pass a similar judgment. Physicians, they argue, "fail to meet commonly accepted and well justified standards for probability interpretation and revision" (Hershey and Baron 1987). Often, Politser (1981) argues, they "equate the probability of a test result, given illness, with the probability of illness given that result. Such misinterpretations are extremely worrisome when it is considered that they may be used as the 'scientific' basis of some major clinical decisions."[33] In a sweeping indictment, Eddy (1982) writes that "physicians do not manage uncertainty very well," that "many physicians make major errors in probabilistic reasoning," and that "these errors threaten the quality of medical care."

It is thus no surprise that physicians can be shown to behave thoroughly idiosyncratically and to treat similar types of patients in astonishingly varying ways. "When people do what comes naturally," notes Lusted (1968), "they are likely to be as inconsistent as can be."[34] Whether seen from the statistical or the information-processing standpoint, physicians are only too human—and humans are "limited in their capacity to think, and biased by cognitive processes that interfere with rational decision making" (Ebbesen and Konecni 1980). Cognitive theories of mental processes can explain these weaknesses. Physicians just lack "the ability to manipulate all these data at once"—the human mind can only hold so many chunks in its short-term memory (Leaper et al. 1972). "There are limits to man's capabilities as an information processor," McDonald (1976) explains, and these limits "assure the occurrence of random errors in his activities" due to (e.g.) "sensory overload."

A redrawing of what medical practice is thus involves a concurrent redescription of what its problems are. Medical practice's "scientific" nature is redescribed, and physicians are subsequently judged in terms of that description. Physicians are seen to "intuitively do statistics" (Clouser 1985), or to generate and test hypotheses—and concurrently it is found that they do so suboptimally. The description of medical action as a scientific activity functions as an explicit yardstick with which it becomes possible to evaluate and criticize what physicians do.

With these new yardsticks in place, new solutions are also called for. Standardizing clinical terminology or introducing problem-oriented records is no longer sufficient. Now that medical work is an activity of individualized minds, the problems of medical practice and their solutions are transformed accordingly. No longer is the lack of awareness of recent scientific developments or the absence of standardized vocabulary or procedures a central matter of concern. These issues are now secondary; what do they really matter when the physician is just not able to make rational decisions?[35] What help is a standardized vocabulary when the complex environment physicians have to deal with simply surpasses their cognitive powers? Different solutions are necessary. "We are trying to solve in our head problems that far exceed the capacity of the unaided human mind," Eddy (1990b) argues in a series of columns in *JAMA*. "There are tools . . . to help us . . . improve the capacity of physicians to make better decisions."

Computer-based decision tools are applauded as the solution to these problems. It is argued that physicians' technical tasks, including diagnosis, the determining of treatment, and prognosis, can be supported by such tools. "The main reason computers are able to conduct these tasks better than physicians," writes Maxmen (1987), "is because the ideal performance of these tasks requires three critical features—memory, objectivity, and probability—qualities which machines display far better than humans."[36]

Likewise, decision-analytic techniques are called for. These, it is claimed, can help the physician since they explicitly take into account matters that are "implicit in every medical decision": "the likelihood of the outcomes of actions, the risks and benefits associated with these outcomes, and value judgments on how the patient's interests are best to be served" (Balla et al. 1989; cf. Silverstein 1988). Since combining and adequately integrating all this information is the physician's weak spot, the power of clinical decision analysis to find the optimal path of action makes it the perfect tool for the job.

Finally, Eddy and others make a case for the protocol. (See, e.g., Komaroff 1982.) In their usage, the protocol fulfills a different role than in the discourse exemplified by the works of Feinstein and Weed. There, the protocol served as a tool to standardize activities over time and space so that the cooperative task of caring for patients could become more efficient—more "scientific." Here, however, protocols are portrayed as the ideal and natural tools for compensating the physician's cognitive incapacities. As Eddy (1990c) eloquently argues, "they are an integral part of our individual and collective psychology." Protocols prevent "mental paralysis or chaos": they "present a powerful solution to the complexity of medical decisions," they "free practitioners from the burden of having to estimate and weigh the pros and cons of each decision," and they "provide an intellectual vehicle through which the profession can distill the lessons of research and clinical experiences and pool the knowledge and preferences of many people into conclusions about appropriate practices" (ibid.). In keeping with the cognitivist discourses, protocols are called upon to support the individual physician's thinking process.

New Practices, New Problems, New Solutions

In the 1982 editorial quoted above, Komaroff said that the "increasing attempts to transform the 'art' of medical decision making into a 'science'" were driven by the "perception that clinicians make medical decisions in an idiosyncratic manner, sometimes compromising the quality of care or wasting medical resources." As we have seen, however, different discourses can be discerned which contain different configurations of what medical practice is, what "scientific" means, what the central problems are, and how these problems have to be dealt with. There has been no gradual transformation of a practice into a science; rather, we have seen a set of discontinuous images in which the notions of "science" and "medical practice" changed shape.

In a similar fashion, Komaroff's remark that this process was driven by a perception of inept decision making cannot be maintained. In the early postwar discourse, I argued, a frontal attack such as Komaroff's on the noble carrier of medicine's art was unheard of. The problems of medical practice were seen to be marginal and to be caused by socioeconomic imbalances that were not the physician's responsibility. A second discourse on medical practice, while framing the relation between the science and the practice of medicine much as the early postwar discourse did, pointed at "interface" problems between science and practice. Sloppy record keeping and terminological inconsistencies obstructed the progress and the utilization of clinical research.

However, the view that the work of physicians themselves was "inept"—that medical practice was internally deficient—arrived only with the discourses that described medical action as a science. Coinciding with increasing attempts by governments, hospital administrations, and insurance companies to get a grasp of the inner workings of health care, medical practice was redescribed in terms of an explicit, normative framework. For Feinstein, Weed, and others, the performance of the "research activity" called medical practice was flawed because physicians often do not adhere to the rules of the scientific method. Owing to the disorganized state of medical record keeping and/or the medical vocabulary, these authors argue, there is no way for physicians to precisely plan, execute, and observe an intervention. Without these basic prerequisites, the scientific method has no leg to stand on.

Thus, not until this discourse were tools to "support" medical action spoken of. Only inside this discourse did a clear-cut niche for the "algorithm" or "protocol" come into being. Now that the distinctive components of the scientific task of diagnosis and treatment are laid bare, and the sloppiness of the actual execution of these components is revealed, the protocol becomes the natural route to adequate performance of the science of medical practice.

Komaroff's observations that "medical decision making" consists of a "spontaneous, informal, and implicit set of judgments," and that "clinicians make medical decisions in an idiosyncratic manner," may be seen as typical of the way medical practice is rephrased in the cognitivist discourses just discussed. In these discourses, medical practice *is* managing mathematical uncertainty, knowing your prior probabilities, or testing hypotheses. Medical practice has become an enterprise of reasoning or calculating minds. In the same vein, the major flaws of medical practice are the limitations of individual cognitive abilities. The new yardsticks applied here, the models of symbolic reasoning or mathematical calculation, reveal a veritable "non-Freudian cognitive psychopathology (Elstein 1982). We have come a long way from the postwar focus on the external, social causes of the problems of medical practice. With the cognitive redefinition of medical practice, the latter had become incapacitated at its very core: the physician's mind.

However, "earlier" depictions can still be found in medical journals and textbooks. The various discourses sometimes mingle, merge, or appear seemingly peacefully side by side. In an introductory chapter to the 1979 *Cecil Textbook of Medicine*, Feinstein criticizes the view that therapeutic action is but an art. Denying that the "purely clinical work of patient care does not present any scientific challenges," he restates his views as portrayed above. In the next chapter of that book, however, J. B. Wyngaarden portrays the relation between the art and the science of medicine much as the early postwar editorials did. The science of medical practice, Wyngaarden (1979) argues, resides in the foundation it provides for the "art of medicine—the skillful application of medical knowledge in the optimal care of the patient."

Notwithstanding these mixtures, the cognitivist discourses on medical practice have become more and more prevalent in medical textbooks and editorials. The 1988 *Cecil Textbook* contains several introductory chapters which stress that medical practice is a scientific activity. "The scientific basis of medical practice is well established, but only recently has the time-honored *art of medicine* come under scientific scrutiny," one of them argues. This "science" is a matter of "intellectual abilities": "collecting information and synthesizing it into integrated concepts compatible with known diseases" (Morgan 1988).

With this growing prevalence of cognitivist discourses, the flawed physician's mind is increasingly seen as a central origin of a broad array of problems of medical practice. More and more, the escalating costs of medical care, the public's dissatisfaction with medical practice, and the suboptimal quality of care are being transformed into problems resulting from the individual physician's mental incapacities.

This transformation affects both the nature and the explanation of the problem. When Komaroff says that the "idiosyncratic manner" of physicians' decision making is "sometimes compromising the quality of care or wasting medical resources," he is not merely attributing a new cause to an already existing problem. In being drawn into the cognitivist discourse, what "quality of care" consists of is transformed: it increasingly becomes a measure of the accuracy of individual decision making. Increasing the quality of care, then, becomes equivalent to "optimizing human cognitive functioning" (Potthoff et al. 1988; cf. Eddy 1990b). Likewise, the problem of the increasing costs of medical care now becomes a problem of "wastage of resources" due to suboptimal decision making of the physician (Eisenberg 1986). Last but not least, in the early postwar discourse the problems of diagnostic error and "unnecessary surgery" were often ascribed to unfavorable circumstances in which physicians were forced to work. Now, however, they are attributed to the occurrence of "pathology" in the physician's decision making.

In line with these changes, decision-support techniques are more and more called upon to aid the physicians' floundering mind, not to amend the structure of medical action. It is to these tools, and the differences between them, that I now turn.

2
Multiple Rationalities: The Different Voices of Decision-Support Techniques

In the 1940s and the 1950s, statistical methods rapidly attained paramount importance in psychology (Gigerenzer and Murray 1987). By the mid 1950s, reporting on experiments without discussing significance levels or null hypotheses had already become unthinkable. This quick reception, Gigerenzer and Murray argue, had everything to do with the desire to create a truly scientific psychology. Incorporating a universal, objective method to test hypotheses was seen as a means of bridging the gap between the foundering attempts of the social sciences and True Sciences such as physics.

As Gigerenzer and Murray point out, the new statistical tools (joined somewhat later with the digital computer) were subsequently transformed into metaphors for the working of the mind. Through looking at the mind in this nonbehavioristic way, new questions could be posed, and new investigational strategies became possible. In studies of perception and detection, for example, the mind was seen to set levels of significance, and to test null hypotheses as to whether stimuli had been perceived. Computers and statistics were scientists' *tools* which were subsequently transformed into scientific *theories* (ibid.).

Analogous transformations have occurred in the case of decision-support techniques. In the previous chapter I demonstrated how the different characterizations of "scientific medical practice" simultaneously constituted different normative frameworks within which the redescribed practice of medicine was judged. By focusing more on the publications of the tools' designers, this chapter will zoom in further on the relations between the discourses and the tools.[1] It will become clear that the normative frameworks, and the distinctive shadows they cast on the physician's performance, have emerged together with the development of the decision-support techniques.

In addition, redirecting the perspective more explicitly to the evolving tools and their builders allows me to elaborate on the theme of the differences between discourses. Through analyzing designers' depictions of their tools and their mutual criticisms, the ideal-typed views of medical practice and "rationality" inscribed in the tools are elucidated.[2] As we zoom in, further differences appear within discourses that previously seemed unified. What seemed to be a single discourse at the level of mainstream medical journals becomes a multitude of fiercely debated, subtly diverging views. With each type of tool, it is demonstrated, come distinct definitions of what medical practice is, what a rational practice looks like, and how a rational practice may be achieved.

I have divided the decision-support techniques into three categories: statistical tools (including diagnostic tools and clinical decision analysis), protocols, and expert systems.

The Quest for Objective Inference: The Statistical Tools

Like the psychologists, the early postwar medical profession also felt that the science of medicine, the footing of medical practice, could benefit from statistical methods. Especially where the design of clinical experiments was concerned, an editor argued, "the clinician should cooperate with the statistician from the start" (Anon. 1952b). Ilana Löwy (forthcoming) argues that this partial surrender of judgmental power should be seen in the light of the increasing public visibility of medicine after World War II. More public funding of both medical care and medical research, she states, meant more political vulnerability for the medical profession. Acquiring objectivity and universality of judgment through statistical methods was a means to strengthen the profession's position relative to a potential critique of idiosyncratic waste of public funds. Further merging the optimistic belief in science with the maintenance of the medical profession's stronghold, then, the promise of an objective, universal means of inferring the truth of a hypothesis from given data was powerful.[3] In 1965 the *Journal of the American Medical Association* started a series called "The Mystic Statistic," the announced purpose of which was to "eliminate the aura of mystery, the suggestion of arcanum, from biostatistics" (Anon. 1965b). Without a sound statistical methodology, it was felt, the research work of clinicians would result in incomparable and falsely drawn conclusions. "Modern medicine, the hard taskmaster of us all" (Anon 1957a) simply demands of us that we "suffer statistics" (Anon. 1965b), editors sighed. More and more, statistics became the *sine qua non* of clinical research.

Thus, an association between the "scientific" character of research and the application of inferential statistics came into being in medicine. As in psychology, and partly through the work of psychologists, statistical theories subsequently started to become models for depicting medical practice itself: physicians were seen as making decisions that were essentially statistical in nature. This phenomenon too was closely linked with coinciding developments in the field of computers. The explorations of what the computer could do for medicine were inseparable from the newly developing views of what medical practice was and how it could be improved. It was felt that "mathematics and mathematical devices can play [a very important role] in biological investigation and in clinical medicine" (Hubbard 1964). The search for this role, however, changed the nature of the game. Builders of statistical tools often cooperated closely with investigators probing the workings of the physician's mind, and they phrased their descriptions of medical practice in the same way (see, e.g., Jacquez 1964 and Jacquez 1972). The statistical techniques and the calculating computer converged as metaphors for the workings of the physician's mind. The intricate intertwining of these developments is portrayed in the first sentences of Robert Ledley and Lee Lusted's influential 1959 article in *Science*:

The purpose of this article is to analyze the complicated reasoning processes inherent in medical diagnosis. The importance of this problem has received recent emphasis by the increasing interest in the use of electronic computers as an aid to medical diagnostic processing. Before computers can be used effectively for such purposes, however, we need to know more about how the physician makes a medical diagnosis.

The answer to this last question, however, was already framed in computational, probabilistic terms: diagnosis had already been modeled to the statistical tool Ledley and Lusted were developing.

Thus, the "tools-to-theory" phenomenon implies that it is only with the coming of these statistical tools that the process of decision making is constantly and matter-of-factly described in statistical terms. Moreover, with different tools, different images arise of what medical practice is and should be. I will discuss two types of statistical tools: *diagnostic tools* (often computer-based) and *clinical decision analysis*.

Statistical Diagnosis

In 1972, a group of physicians and computer scientists led by F. T. de Dombal published a series of articles on a "computer-aided diagnosis" system running in the Department of Surgery at the University of Leeds. The

Abdominal Pain Chart

NAME		REG NUMBER	
MALE/ FEMALE AGE		FORM FILLED BY	
PRESENTATION (999, GP, etc)		DATE	TIME

PAIN

SITE	AGGRAVATING FACTORS	PROGRESS
	movement	better
	coughing	same
ONSET	respiration	worse
	food	DURATION
	other	
	none	
		TYPE
	RELIEVING FACTORS	intermittent
	lying still	steady
PRESENT	vomiting	colicky
	antacids	
	food	SEVERITY
	other	moderate
RADIATION	none	severe

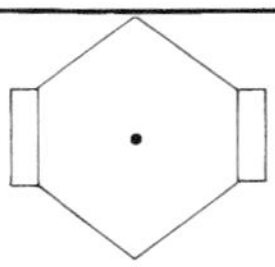

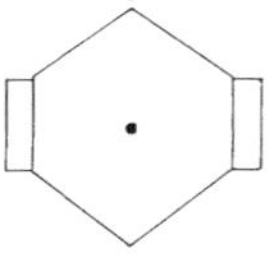

HISTORY

NAUSEA yes no	BOWELS normal	PREV SIMILAR PAIN yes no
VOMITING yes no	constipation / diarrhoea / blood / mucus	PREV ABDO SURGERY yes no
ANOREXIA yes no	MICTURITION	DRUGS FOR ABDO PAIN yes no
PREV INDIGESTION yes no	normal / frequency / dysuria	♀ LMP pregnant
JAUNDICE yes no	dark haematuria	Vag. discharge dizzy/faint

EXAMINATION

MOOD normal distressed anxious	TENDERNESS REBOUND yes no	INITIAL DIAGNOSIS & PLAN
SHOCKED yes no	GUARDING yes no	RESULTS
COLOUR normal pale flushed jaundiced cyanosed	RIGIDITY yes no MASS yes no	amylase blood count (WBC) computer urine X-ray
TEMP PULSE BP	MURPHY'S +ve −ve	other
ABDO MOVEMENT normal poor/nil peristalsis	BOWEL SOUNDS normal absent +++	DIAG & PLAN AFTER INVEST
SCAR yes no	RECTAL — VAGINAL TENDERNESS left right general mass none	(time)
DISTENSION yes no		DISCHARGE DIAGNOSIS

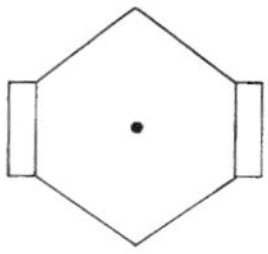

History and examination of other systems on separate case notes

Figure 2.1
A form used with de Dombal's tool (de Dombal 1991). Copyright 1991 Société d'Édition de l'Association d'Enseignement Médical des Hôpitaux de Paris. Reproduced by permission.

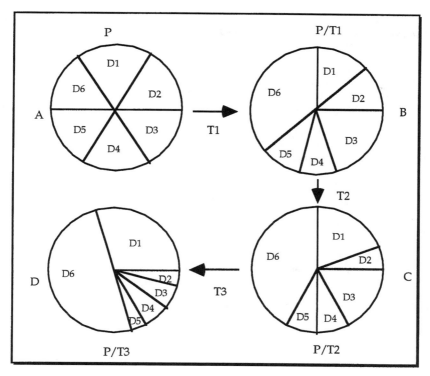

Figure 2.2
Bayes' Theorem states that one can calculate the probability that a member of a population having a symptom has a disease by means of the formula P(D|S) = (P(S|D) × P(D))/P(S). In words: once you have the probability of exhibiting the disease (P(D)), the probability of having the symptom (P(S)), and the probability of having the symptom if one has the disease (P(S|D)), you can calculate the chance that a member of your population with symptoms S has disease D (P(D|S)). This illustration (source: Sultan et al. 1988) shows how the sequential application of this formula gives a disease (D6) prominence over five alternative diagnostic categories. Each sequential application (T1–T3) incorporates an additional symptom or test result (S). Copyright 1988 *Hematologic Pathology*. Reproduced by permission.

system dealt with the diagnosis of appendicitis, which in the absence of suitably specific tests is a notorious problem in emergency medicine. Upon completion of a form listing the required historical and examination data (figure 2.1), a computer utilized a statistical formula (Bayes' Theorem) to calculate the probability of a certain diagnosis for the patient in question (figure 2.2). The knowledge base of this system consisted of previously entered data on actual patients; the diagnoses the computer

```
POSSIBLE DIAGNOSIS
APPEND DIVERT PERFDU NONSAP CHOLEC SMBOBT PANCRE
PROBABILITIES ARE
   0.0    0.0    2.7    0.0    0.9    3.1    93.2

CLINICIANS DIAGNOSIS
PRIMARY      -CHOLEC
SECONDARY   -SMBOBT

COMPUTERS DIAGNOSIS
PRIMARY      -PANCRE 93.2
SECONDARY   -SMBOBT 3.1

NEITHER OF YOUR DIAGNOSES SEEM LIKELY. PROBABILITIES
INDICATE PANCRE AS PRIME POSSIBILITY

++ SUGGEST CHECKING THE FOLLOWING.......
AMYLASE
TENDERNESS....
SITE PRESENT
```

Figure 2.3
A printout from de Dombal's system (source: Horrocks et al. 1972). The computer has compared its diagnosis to the "clinician's diagnosis," which can be optionally typed in along with the clinical data. The seven mentioned diagnoses are acute appendicitis, acute diverticular disease, perforated duodenal ulcer, non-specific abdominal pain, acute cholecystitis, acute obstruction of the small bowel, and acute pancreatitis. Copyright 1974 BMJ Publishing Group. Reproduced by permission.

could pick from represented the seven diseases most often causing acute abdominal pain, together covering more than 95 percent of all cases.

According to Horrocks et al. (1972), the whole process of typing, processing, and printing the results (see figure 2.3) seldom took more than 20 minutes and often took no more than 5. The system performed well: de Dombal and his colleagues claimed that the computer had a diagnostic accuracy of more than 90 percent, while experienced physicians hardly ever reached 80 percent. Dramatically, the authors added that this implied that use of the tool would significantly decrease the rate of "unnecessary" operations. Fewer healthy appendixes would be removed, without inducing an increase in the number of ruptures due to delaying surgery. Some years later, their trials showed that this was indeed the case (Adams et al. 1986; de Dombal 1989).

Many tools like de Dombal's computer-aided system were built in the 1960s and in the early 1970s: tools for diagnosis of congenital heart disease, thyroid disease, and so forth.[4] The builders of these tools describe medical work in specific ways. De Dombal and his team see physicians, like their tool, as making diagnostic decisions by "working from a set of . . . probabilities" (Leaper et al. 1972). Medical diagnosis is a "problem in conditional probability" (Anon. 1961a); the physician is a "diagnostic computer," "turning out possibilities on the basis of concurrence of several manifestations, together with such factors as age, sex, and the sequence of events" (McDermott 1971). Consciously, intuitively, or otherwise, the physician comes to a diagnosis through a process of statistical inference.[5]

As argued in the previous chapter, once the diagnostic task is seen as an exercise in statistics, physicians' performance is seen as second-rate. In this view, physicians, like everyone else, "are not able to extract from data nearly as much certainty as is latent in the data" (Lusted 1968; see also Cebul 1988). The computer can do this for them. Since differential diagnosis is now a problem in conditional probability, it "may be solved with a computer with an accuracy which depends only upon the accuracy of the statistical data regarding the incidence of symptoms in diseases and the accuracy of the data collected from the patient in question" (Anon. 1961a). Moreover, by continually updating its database, the computer can benefit better from its "experience" than the physician can (ibid.).

Thus, the tool provides the normative yardstick against which the physician's performance is judged. If the physician's vague, inadequate diagnostic "judgment" were to be supplemented by the computer's objective calculations, it is argued, medical practice would be better off.

What is the rationality inscribed in these techniques? Rational decision making here implies the existence of a formula that automatically reaches the best possible diagnosis when given a set of data. Context is deemed irrelevant, since the essential elements of the decision are caught in the formula. All other context is bias. Content would ideally be irrelevant too. According to de Dombal's team, the ultimate ambition is to have the computer, given a set of data, select from its files "the most appropriate diagnosis from the whole spectrum of recognized clinical ailments" (Horrocks et al. 1972). Ideally, the statistical tool performs the trick, regardless of the specifics of the problem at hand.

Medical practice is said to be made more rational by replacing the physician at the supreme moment—that of the diagnostic decision—with a process of mechanized statistical inference: "The computer's conclusion will not be biased by irrelevant factors. For instance, the computer will not weigh most heavily its most recent experience, as a physician is prone to do. The computer performs just as well at 2 A.M. as at noon."

(Anon. 1961a) Statistical tools, which had become the *sine qua non* of clinical research and had turned the clinician's experimenting into a truly scientific activity, now also became the epitome of rational medical practice (cf. Gigerenzer et al. 1989).

Clinical Decision Analysis: Reshaping the Statistical Tool

The development of clinical decision analysis, which gained momentum in the 1970s, can be seen as an attempt to broaden the scope of the diagnostic tools. The method its advocates championed dealt with making optimally rational choices in situations of uncertainty. It grew out of mathematical approaches to decision problems that had gained much interest and impact through their usage by the military during and after World War II.[6] Decision analysis merges Bayesian statistics with economists' utility theory: it opts to select the action with the highest expected utility from a range of possible actions. Tying into the increasing public problematization of health care expenditures, then, the decision analysts' focus on "costs" and "benefits" and the economic model of rational man aligned with the "economization" of the medical profession's vocabulary.

The steps of clinical decision analysis were drawn directly from the original work of decision analysts such as Leonard Savage (1954) and Howard Raiffa (1968).[7] To find the action with the highest expected utility, one multiplies the utility of each possible individual outcome (i.e., the value of each outcome for the patient, expressed as a number) by the probability that each outcome will occur. The action that yields the highest total expected utility is the action of choice (figure 2.4). When the possible outcomes and intermediary decisions increase, the tree gets very complex (figure 2.5). Nevertheless, the method should in principle be able to generate the optimal decision for any medical decision problem.

Both decision analysts and builders of diagnostic tools see the physician's decisions as quantitative operations.[8] For decision analysts, however, to focus primarily on diagnostic decisions is to miss the important ones. In medical practice, they argue, the important decisions are not so much those that concern what is the case as those that concern what to do. In other words, what matters in medical practice is the *selection of actions*—including diagnostic actions. "Diagnosis" is just one little step in the whole process of "patient management" (Ginsberg 1972), whereas "action-selection" is of continual importance (Lusted 1968). Moreover, decisions about which diagnostic test or therapeutic procedure should be used often have to be made whether or not a diagnosis is already known. In fact, decision analysts argue, a "diagnostic decision" as such is hardly

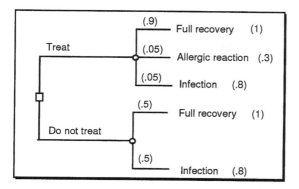

Figure 2.4
A simple decision tree for throat infection. A square node is a decision point; circular nodes indicate chance occurrences. The decision analyzed is "to treat or not to treat." Possible outcomes are listed, and each possible outcome is assigned a probability (0.9, 0.05, and 0.05 for the upper branch) and a value, between 0 and 1, stating its utility (1, 0.3, and 0.8 for the upper branch: an allergic reaction, which might be fatal, is worse than an infection, which is usually self-limiting but might lead to complications such as peritonsillar abscesses). The net score of "treat" is $(0.9 \times 1) + (0.05 \times 0.3) + (0.05 \times 0.8) = 0.955$, and the net score of "not treat" is $(0.5 \times 1) + (0.5 \times 0.8) = 0.54$. Since the expected maximal value of "treat" is higher than that of "do not treat," the physician—given this situation, these probabilities and these utilities—should treat.

ever taken. Most of the decisions that physicians face are like the decision trees portrayed in figures 2.4 and 2.5, where treatment is required while the diagnosis is not yet known (figure 2.4) or where a decision has to be made between a diagnostic action and a therapeutic action (figure 2.5).

Selecting an action in an uncertain situation, moreover, is always more than just factually knowing what is the case. In all decisions, these authors argue, *value judgments* have to be taken into account. A merely diagnostic approach "has the serious deficiency that it is indifferent to the risks and pain involved in various tests and has no way of balancing the dangers and discomforts of a procedure against the value of the information to be gained. In this sense it lacks a key element that characterizes the practice of a good physician." (Gorry et al. 1973) Choosing an action thus implies weighing risks and benefits—something the diagnostic tools could not do. Clinical decision analysis, according to these authors, offers an integrated, quantitative, objective approach to all physicians' decisions, combining the probabilistic nature of medical practice with the need to take the utility of outcomes into account.[9]

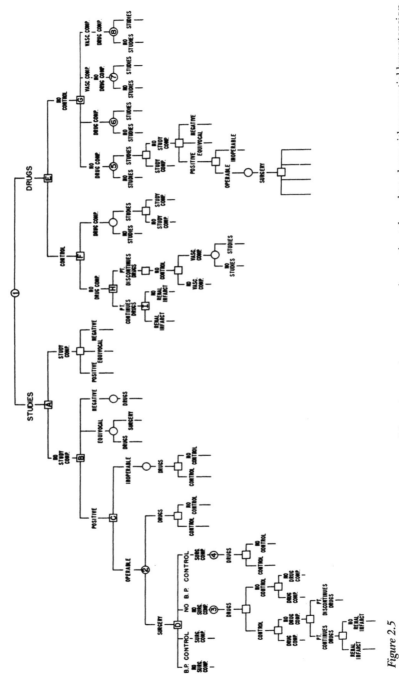

Figure 2.5
A more complex decision tree, describing possible actions and consequences in a patient thought to have either essential hypertension or renal artery stenosis. Source: Schwartz et al. 1973. Copyright 1973 Excerpta Medica Inc. Reproduced by permission.

Another major difference is that decision analysts see the *patient's perspective* as an essential part of the rationalization of medical practice. Coinciding with a general trend in this period (Armstrong 1983; Arney and Bergen 1984), clinical decision analysis discovered "the patient as subject." Up to this point, Slack (1972) argues, the physician was expected to make the decisions for the patient, "all on a 'doctor-knows-best' basis." Medicine, however, should become "patient centered" as much as Carl Rogers's client-centered therapy: "[There should be a kind of] partnership with the patient. In a step-by-step manner, advancing through the diagnostic process, the physician will communicate possible plans of action, and the patient will decide what should be done." (ibid.) For the diagnostic tools, the patient existed only as an object, from which clues had to be gathered to come to the right diagnosis. Here, however, the patient's subjective preferences (incorporated as "utilities") are a fundamental input of the decision-analytic procedure. Adequate rationalization of medical practice, these authors hold, is impossible without full and systematic attention to the patient's voice (Schwartz et al. 1973; Balla et al. 1989).

The final major modification introduced by the founders of decision analysis was the usage of *subjective* probabilities. The early diagnostic tools discussed in the previous paragraph, such as de Dombal's, used objective probabilities, which are generated by analyses of large samples of patients. A major shortcoming of objective probabilities, however, is that the requisite data are often hard to obtain. Since the scope of clinical decision analysis is so much broader, this problem is even more poignant in that context: "When seeking the data for test characteristics, therapeutic effectiveness, risk, and cost, the clinician hoping to use decision analysis is confronted with a bewildering array of conflicting definitions, methodologies, and means of reporting findings. Often, related research has not asked the decision-oriented questions that are of interest to the clinician, and it therefore does not even purport to supply the relevant data." (Cebul 1984)

When one cannot find the "massive amounts of clinical data" (Gorry et al. 1973) needed to build a decision tree in the medical literature, one can of course try to elicit the required probabilities from the files of one's own patients (as de Dombal's team did). The creation of a "sufficiently large and reliable database," however, is a "slow, tiring and troublesome job" that can easily take years (Bjerregaard et al. 1976).

To overcome this, decision analysts recommended using subjective probabilities, or "disciplined personal opinion" (Savage 1972). To keep the rationalization of medical practice from getting stuck at the outset, the physician is asked to supply what he or she thinks, on the basis of personal experience, are the probabilities. Of course, decision analysts admit, these probabilities are often rather inaccurate. However, "when there is a paucity of objective evidence at hand, we require a methodology that brings information, however vague and imprecise, into the analysis, rather than a method that suppresses information in the name of scientific objectivity" (Raiffa 1968). The power of statistical tools should not be left to wither in those tiny sections of the realm of medicine where objective probabilities do happen to be available.

Moreover, the absence of proper data is a problem for "our traditional, implicit or tacit clinical decision making" (Kassirer et al. 1987), decision analysts contend. "Because a decision must be made, the question is not whether probabilities will, in fact, be used, but what estimates are to be employed." (Schwartz 1979) Using the formal method of decision analysis, then, can only improve the quality of the decisions made.

Different Tools, Different Practices, Different Rationalities
The dissimilarities between clinical decision analysis and the diagnostic tools reflect different depictions of medical practice and different rationalities. Many builders of diagnostic tools do not warmheartedly agree with the decision analysts' alternatives. They fear that the increase in applicability and scope brought by decision analysis is bought at the price of the scientific, objective nature of their tools. The ideal of objective inference, in the eyes of some, is invalidated too much. For one thing, the use of subjective probabilities is debated intensely. Many statisticians do not even accept these probabilities as meaningful in the first place. Does it make any sense, these critics ask, to multiply a subjective probability by another? Or to add them (cf. Schoemaker 1982)? Moreover, cognitive psychology has shown that humans are rather poor at estimating probabilities. As Tversky and Kahnemann (1974) have argued, subjective probabilities show "large and systematic departures from proper calibration" and tend to change when the procedure by which the estimate is elicited is altered. What, retort the diagnostic tool builders, should we make of decision trees construed with such poor information? When de Dombal's tool was programmed with subjective probabilities obtained from physicians, it performed almost as poorly as the physicians themselves (Leaper 1972; interview with de Dombal, 1993). Are we not creating "silk purses from sows' ears" (Cebul 1984)? What is left of the ideal of objective inference when all this subjectivity is let in through the back door?

Likewise, the emancipatory, humanistic ideal of incorporating patients' preferences is regarded with skepticism. The gathering of utilities is controversial and strewn with problems, decision analysts themselves admit. Elaborate and lengthy modes of questioning are needed to gather them properly, involving questions like "Would you rather have a 10 percent chance of death but not lose your leg or have your leg amputated and not run a chance of dying?" (Cf. Weinstein et al. 1980.) Since these methods are so awkward, some authors argue, your measurements may tell you more about the method used than about the patients' actual preferences (Hershey and Baron 1987). Utilities appear to change when questions are framed differently and are highly unstable over time (Kassirer et al. 1987; see also Politser 1981).[10] Gathering utilities is the "art and science of decision analysis" (Bell et al. 1988). Here again, it is argued, much vagueness and subjective judgment are let in.

The same can be said of the creation of the tree. The difficulty of this process is often commented upon. What outcomes should one include? In the example given in figure 2.4, should additional side effects of the antibiotics be included? Should the branch "do not treat" be expanded by including an option to wait one day to see whether the infection subsides—and if not, then treat? How detailed should one get? How important are omissions? These choices, decision analysts often state, are also part of the "art," the "subjective part" of decision analysis, where error lurks. The "fidelity" of the trees is often poor, Kassirer (1987) and his co-workers remark; often, important outcomes are simply overlooked (Schwartz 1979).[11] The use of the tree itself, in other words, already violates the ideal of objective inference. The way it is structured greatly influences the outcome of the decision analysis, yet it is saturated with subjective guesses and unquantified *content*.[12]

So we see two different tools, linked to two different views of what medical practice is, how it should be rationalized, and what "being rationalized" means. The advocates of these tools disagree about what the central "act" of medical practice is, and about the ideal of objective inference. Although both groups of authors adhere to the same ideal, clinical decision analysts feel that it should be interpreted somewhat liberally in order to get a meaningful tool in the first place. In contrast, diagnostic-tool builders point out that by condoning this polluting of the ideal of objective inference one is only replacing one type of messiness with another.[13]

We have seen how two types of tools lay at the root of the "statistical discourse" described in the previous chapter. The differences between these tools are hard to discern from the perspective of the mainstream medical literature; they come into view only when one closes in on the specific tools. Only then does it become clear how, in and through the development of these tools, medical practice has been redescribed in their likeness. Only with the coming of the statistical diagnostic tools was the process of differential diagnosis seen as a problem of conditional probabilities. The same process occurred in the case of clinical decision analysis, which is often said to do nothing but "make the hidden decisions explicit" (Lusted 1975). This did not become the case, however, until the desires of patients were seen as the expression of "personal utility structures" and medical decisions were redescribed as "weighing risks and benefits" (Weinstein et al. 1980)—transformations which only occurred with the coming of decision analysis.

Capturing the Clinic: The Protocol

A second category of tools drawn upon to rationalize the practice of medicine includes an array of techniques which go by a plethora of names: guidelines, algorithms, practice policies, standards, statements, protocols. The various labels are used in widely divergent ways: what Eddy calls "practice policies," the US Committee on Clinical Practice Guidelines calls "guidelines"; what some call "protocols," others refer to as "algorithms."[14] All these tools, however, have in common that they are or can be read as a set of *instructions* telling medical personnel to do a certain thing in a certain situation. These instructions may be more or less elaborate, precise, or binding; they may be formatted in different ways, and may or may not be construed in a structured way—but they all share this feature. They may be elaborately designed as a flow chart containing great detail, or they may consist of a number of rather vague and general recommendations, but they all guide medical personnel through a sequence of steps. It is as an umbrella term indicating this feature that I use the term "protocol."

In view of this plethora of terms and techniques, it will come as no surprise that the history of the protocol in medical practice is difficult to circumscribe. The emergence of such tools in medical practice was much less of a traceable, concentrated activity than the development of the statistical tools. In its simplest form, a protocol is nothing but a written instruction; and, of course, the idea of regulating action through a recipe is an ancient one (Goody 1977). Protocols figured in most of the discourses discussed in chapter 1, including protocols intended to overcome troubles generated by agencies external to medical practice (fitting the early postwar discourse on medical practice and its problems) and protocols intended to compensate for the physician's cognitive incapacities. In the latter case, the resemblance of the sequential and logical nature of the protocol to mental processes was emphasized.

It cannot be said, then, that "the protocol" was uniquely tied to one of the discourses on medical practice discussed in chapter 1; it has been incorporated in several of them. On the other hand, I do argue for a special link of the protocol to the discourse Feinstein and Weed helped come into being. It was in this discourse that protocols, as a tool relevant for medical practice, came into full view—and in a highly specific way.

Feinstein and Weed modeled the practice of medicine to the steps of the scientific method, as taught by neo-positivist philosophy of science. Riding the wave of standardization efforts that was washing over the practice of medicine, Weed and Feinstein argued that standardization could do for medical practice what it had done for medical science. The success of medical science during and after World War II had shown the merits of strengthening the collective effort through the coordination and linking of individual actions. Protocols were part and parcel of this enterprise: in the booming field of clinical research, the protocol was essential to ensuring that the actions and interpretations of outcomes would be similar in all participating institutions (Howard 1961; Marks 1988).[15] If medical practice would follow—if cooperative work in this particular science would also be made more efficient and communication would be streamlined—it too would flourish. Feinstein and Weed brought the protocol and "(scientific) medical practice" together in a way that transformed the meanings of both notions. Medical practice became defined as the logical and sequential (i.e., protocol-like) execution of the steps of the scientific method; the standardizing protocol, concurrently, became a tool with which to structure and coordinate the scientific work of diagnosing and treating patients.

The circumscribed view of medical practice and rationality connected to this protocol and the way it is positioned vis-à-vis the other types of tools are discussed in the remainder of this section. I first discuss Feinstein's eloquent and harsh criticism of the statistical tools. Next I describe a typical example of the type of tool that is linked to this kind of criticism: a protocol for "physician extenders" developed in the early 1970s. As can be expected, the builders of statistical tools regarded the criticism and the tools of Feinstein and others with skepticism; their countercritique is discussed in the last subsection.

The Haze of Bayes and the Aerial Palaces of Decision Analysis: A Criticism of Statistical Reasoning[16]

Decision analysis has led us into Vietnam: where will it lead health care?
—*graffito, Harvard School of Public Health, 1979*[17]

Feinstein attacked both the practical functioning of the statistical tools and their inscribed ideal-typed rationalities and views of medical practice. In his view, medical practice is nothing like a mathematical activity, and the ideal of objective statistical inference is an utterly mistaken goal. His criticism, developed in many years of prolific writing, can be somewhat crudely summarized in two main points.[18]

First, *why replace judgmental decisions with a logic of statistical reasoning when the latter requires judgmental decisions too?* The statisticians' critique of clinical judgment, Feinstein (1987a) argues, is that it "often may be applied in an inconsistent, unstandardized, or even capricious manner. The great appeal of ... mathematical models is that they offer a standardized mechanism for this process." The problem with this endeavor, Feinstein states, is that it *also* requires many *ad hoc*, nonmathematical judgments. First of all, mathematical tools always imply a wide range of assumptions. Medical practice, however, more often than not violates these assumptions:

To apply the Bayesian concept in general clinical diagnosis requires assumptions about nature that are incompatible with realistic clinical activities. The types of data and the required "independence" of different variables in the Bayesian diagnostic calculations can be obtained only if clinicians ignore the epidemiologic realities of human ailments: many people have ... [undiscovered] disease, and many people have multiple co-existing diseases. (Feinstein 1967)[19]

When are such violations severe? When are they allowable? Questions like these continually require pragmatic, subjective judgments. Clinical decision analysis faces such questions even more frequently: as the statistical tool builders already charged, setting up the tree, estimating subjective probabilities, and eliciting utilities involves continual judgmental decisions (cf. Feinstein 1987b).

Stressing the fact that he is a clinician himself, Feinstein (1977) argues that statisticians hardly understand how inevitable and frequent these violations will be. The logic of statistical reasoning will simply drown in the messy reality of actual clinical practice: it completely overlooks the pragmatic, fluid nature of clinical reality. Statistical tools, for example, often require the input of sensitivity and specificity values of the relevant diagnostic tests (measures of the capacity of the test to distinguish between the absence or presence of e.g. a disease). "The mathematical quantophrenia," however, leads us to overlook that the sensitivity and specificity of a test are often highly dependent on the reason *why* a physician requests the test (whether for discovery, exclusion, or confirmation), the severity of the disorder, and so forth. In different situations, we are dealing with different groups of patients, which thoroughly affects the magnitude of these measures. When all these matters are taken into account, Feinstein (1985) asserts, "we can promptly begin to discern the scientific inadequacy of all the mathematical folderol."[20]

Similarly, diagnostic and therapeutic actions are often interwoven in medical practice, making a Bayesian "diagnosis-machine" irrelevant (Feinstein 1973a). Here, decision analysis is not the solution it claims to be, since building decision trees demands the utterly unrealistic attempt to foresee all possible outcomes and all intermediate actions and to decide upon all the required probabilities and utilities (Feinstein 1977). The pointlessness of this task is even more glaring when one considers the building blocks used. Medical nosology is a "potpourri," Feinstein continually insists, with ambivalent terminology and vague criteria that mean something different to every other physician. Any attempt to build a clean, precise, mechanized inference tool upon this quagmire will be thoroughly thwarted from the outset (Feinstein 1973a).

Feinstein (1987a) argues, thus, that "the use of the mathematical models merely transfers the judgmental decisions from one intellectual location to another." The messiness of medical practice is much more widespread and unyielding than the statistical tool builders appear to think. As a consequence, these decisions are transferred into the hands of people who do not know anything of the intricacies of medical work: statisticians. The tools we end up with, Feinstein concludes, are nothing but "a splendid array of castles in the air—having all the abstract academic virtues of aerial palaces and none of the gritty dirt associated with a strong foundation or firm roots in reality" (Feinstein 1977).

Feinstein's second main point, closely related to the first, is that *statistical goals often conflict with clinical goals.* "Soft" data (such as visual impressions or psychosocial information), Feinstein argues, are of crucial importance in medical practice. Such data invariably tend to drop out the statistical tools as "too vague" (Feinstein 1967), thereby threatening to further dehumanize medical practice. (See figure 2.6.) Moreover, and crucially, Feinstein (1987a) argues that the "decontentualized" nature of the statistical tools is unacceptable and utterly unscientific in a clinical context:

Bayes's rule of inverse probability.

Figure 2.6
A cartoon criticizing a Bayesian approach to medical practice. Source: van Bemmel and Willems 1989 (no original reference). Copyright 1989 Bohn, Scheltema & Holkema. Reproduced by permission.

The mathematical goals are aimed at eliminating details, using standardized models, and producing maximum reductions of variance in the available data. . . . [If] the clinician wants to preserve details, observe direct evidence of relationships, . . . and arrive at conclusions that are clinically both cogent and consistent, the conventional mathematical goals will not always be satisfactory.

Physicians do not simply want to hear that a symptom is correlated with a disease. They want to *understand.* The essence of scientific action, after all, is *explaining* this correlation through (for example) a pathophysiological mechanism. Statistical methods merely "label," Feinstein (1973a) argues. Their conclusions "may be accurate but impertinent, . . . statistically satisfactory but scientifically defective, diagnostically inadequate, and therapeutically hazardous" (Feinstein 1973b). More often than not, clinicians are not very interested in the "number issued by the calculations of probability" (Feinstein 1967); their actions are guided by "many features much more subtle than a simple calculation of diagnostic probabilities" or a "clinically alien arrangement of statistical scores"(Feinstein 1987b).[21]

Feinstein (1973b) stresses that he is by no means an "antistatistical nihilist." It is just the "unscientific manner" (1994) in which statistics are being applied to medical practice which bothers him. Developing good tools, Feinstein (1973a) argues, requires "adapting statistics and computers to the practical realities of clinical medicine rather than forcing clinical phenomena into Procrustean modifications to fit the theoretical concepts of statistics and computers."

The Clinical Grounding of the Protocol

In 1974, Greenfield et al. reported on their experience with the use of a protocol for "physician-extenders."[22] Patients with upper-respiratory-tract complaints entering the walk-in clinic of Beth Israel Hospital in Boston were seen by a health assistant—usually a briefly trained high school graduate. The health assistant first scanned a list of "chief complaints" indicating suitability for the protocol (figure 2.7). If one of these complaints was present, a data-collection form (figure 2.8) guided the health assistant through a series of questions regarding symptoms and relevant history (did the patient "ache all over," was the patient taking antibiotics, and so forth). Furthermore, the protocol required some signs to be assessed through physical examination (e.g., whether the sinuses or the neck nodes were tender). Finally, the health assistant followed the steps outlined in the "decision-making algorithm" (figure 2.9). The algorithm checked for the absence or presence of sets of symptoms and selected one out of four possible "medical actions": send home without specific treatment, make throat culture, give penicillin, or refer to the physician.

Acute cough (<one-week duration)
'Cold'
'Flu' or influenza
Hay fever
Hoarseness
Postnasal drip
Sinus trouble
Sneezing
Sore throat
Streptococcal throat infection
Stuffy or runny nose
Tonsil trouble or tonsillitis
Request for throat culture

Figure 2.7
A list of chief complaints, indicating suitability for a protocol. Source: Greenfield et al. 1974. Copyright 1974 American Medical Association. Reproduced by permission.

58   Chapter 2

```
           Name _____  Age _____  Sex _____  Date _____
           Chief complaint _____

        Y   N                              Subjective
       ┌──┬──┐
       │  │  │    Sore throat: duration _____
Y─▶    ├──┼──┤
R─▶    │  │  │    Cough: duration _____
       ├──┼──┤
       │  │  │         production, substantial or increased
       │  │  │         chest pain
R─▶    ├──┼──┤
Y─▶    │  │  │         even when not coughing
       ├──┼──┤    Runny/stuffy nose: duration _____
B─▶    │  │  │         more than 3 times a year
B─▶    │  │  │         tearing with stuffy nose
B─▶    │  │  │         itchy eyes with stuffy nose
B─▶    │  │  │         itchy nose with stuffy nose
B─▶    │  │  │         attacks of sneezing with stuffy nose
B─▶    │  │  │         allergy in parents or siblings (not self)
       ├──┼──┤    Ear ache: duration _____
R─▶    │  │  │         with discharge
R─▶    │  │  │         with hearing impairment
R─▶    │  │  │         pain even when not swallowing
Y─▶    ├──┼──┤    Hoarseness (by observation): duration _____
R─▶    │  │  │    New skin rash
       │  │  │    Headaches
R─▶    │  │  │         severe (restrict normal activities)
Y─▶    │  │  │    Ache all over
R─▶    │  │  │    History of rheumatic fever
       │  │  │    Exposure to strep in past week
       │  │  │    History of penicillin reaction
R─▶    │  │  │    Return visit for same complaint _____
       └──┴──┘
                                              Rx
       ┌──┬──┐
R─▶    │  │  │    Taking antibiotics
       ├──┼──┤    Taking anything for chief complaint _____
       │  │  │         does it work
       └──┴──┘
                                           Objective
       ┌──┬──┐
R─▶    │  │  │    Tender sinuses
       ├──┼──┤    Tender neck nodes
R─▶    │  │  │    Lip/mouth sores                    Impression: _____
       ├──┼──┤    Exudate                            Plan: _____
       └──┴──┘    Temperature _____          _____
                                                     _____
       ┌─────────────┐                                _____
       │ Code        │                               Signature _____
       │ Y = yellow  │
       │ R = red     │
       │ B = blue    │
       └─────────────┘
```

Figure 2.8
A data-collection form for a protocol (Greenfield et al. 1974). Information is denoted in the columns as present (Y) or absent (N). R, Y, and B in the rows indicate the colors (red, yellow, and blue) of the boxes used in the decision-making algorithm. Copyright 1974 American Medical Association. Reproduced by permission.

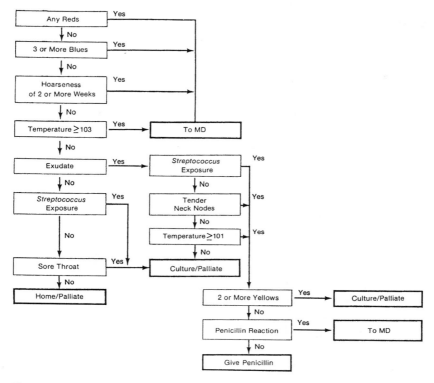

Figure 2.9
A decision-making algorithm of the Greenfield protocol (ibid.). Copyright 1974 American Medical Association. Reproduced by permission.

The major "medical reasons for accurate work-up of upper respiratory tract infection," Greenfield et al. (1974) argue, "is the detection of streptococcal pharyngitis": a bacterial infection of the throat for which antibiotic treatment is effective. Moreover, "if the patient does have a streptococcal infection, he might be at higher risk for rheumatic fever."[23] The protocol attempted to ensure that patients who might have something "serious," or something requiring a different diagnostic workup, would immediately be sent to a physician. One symptom or sign marked "red" or three or more symptoms and signs marked "blue" (figure 2.8) were taken to indicate that something like pneumonia, herpes infection, otitis media, or an allergy might be present.

The algorithm makes "a differential diagnosis based on the assumption that clinical judgment flows from an analysis of many overlapping features that may be present to varying degrees" (ibid.) In this case, the challenge is that the symptoms and signs of a viral throat infection (indicated by the presence of more than two yellow symptoms) may appear similar to those of a streptococcal infection. "In choosing what attributes are to be included for directing the protocol logic," Greenfield et al. continue, "decisions are derived from a combination of practical and purely medical considerations, [which were subjected to peer and consultant review]. These may not be universally acceptable to all physicians; however, if they are explicit, each can be rejected or modified to fit the view of the individual physician." So, Greenfield et al. argue, "under ideal circumstances" one would probably never treat immediately with penicillin: the diagnosis would first have to be confirmed through a positive culture. Only then would the patient be treated. However, they state, "lost cultures, lost patients, and a respectable rate of false-negative single cultures (about 10 percent) make waiting hazardous. For these reasons we treat these patients immediately, in accordance with the practice of many, and in what we believe is a sound convergence of theoretical and practical considerations." A culture is taken only if the algorithm ends up with a clinical picture with "overlapping features": when a distinction between viral or streptococcal infection cannot be made.

Greenfield et al. tested the protocol by having it administered to 226 patients, who were subsequently seen by a physician for evaluation. All in all, they say, 42 percent of the patients would have been sent home by the protocol—of which none, according to the physicians, had pathological conditions other than uncomplicated (nonstreptococcal) upper-respiratory-tract infections. This would have meant a considerable savings of physicians' time. Furthermore, Greenfield et al. argue, the protocol is safe, since "the decisions about data-base collection and disposition are not made by the health assistant. They are made by the protocol, derived from local experience and peer consensus."

What is the image of medical practice associated with these types of protocols?[24] What is a rational practice, and how is such a practice to be achieved? Good algorithms, Feinstein (1974) argues, "describe good clinical reasoning: rules that are specific enough to manage the standard situations, broad enough to encompass the common exceptions, and flexible enough to allow separate decisions for the rare." The protocol structures medical practice in such a way that its essential clinical nature is retained, while the "reproducibility and standardization" that will make a true science out of medical practice are introduced. "The object is to preserve the vitality of clinical reasoning while enhancing its scientific effectiveness." (ibid.) Thus, the protocols contain pragmatic but optimal, sequential paths through data gathering, interpretation, requesting additional paraclinical tests, prognostic estimations, therapeutic actions, and behavioral strategies. As in Greenfield's protocol, pathophysiological theories are and have to be connected to practical considerations, and diagnostic decisions often come only after the therapeutic act. Protocols try to strengthen this clinical rationality, without trying to replace it with some alleged superior rationality (as the statistical tools did). A protocol simulates the physician's actions but includes "the rationale for each decision which it makes" (Komaroff et al. 1973). As Feinstein (1974) argues, "the addition of suitable justification, containing citations of data or principles to substantiate each decision, is the activity that converts a flow chart [a term for the representation of a protocol as in figure 2.9] from an arbitrary set of rules into a scientific document."

"Rationality," thus, coincides here with attention to "soft" data, with the need to understand why (against the statisticians ideal of statistical inference), and with a pragmatic stand forced upon one by the demands of the clinic.[25] In addition, the ideal-typed protocol intervenes in a different way than statistical tools. The latter, in principle, focus only on the moment of decision. At that moment, when physicians are at their weakest, the tool momentarily takes over. In contrast, the protocol standardizes a whole sequence of actions. A protocol can contain several decisions at once, since "an efficient clinician regularly intermingles many other decisions with the diagnostic reasoning" (Feinstein 1974; see also Margolis 1983)—such as prognostic estimations, choices of additional tests, selection of therapeutic agents, and planning for the personal interaction with the patient. Although many current protocols do have a single, often diagnostic decision as their end point (as does Greenfield's protocol), the realities of clinical work will require an increase in the production of protocols as exemplified in figure 2.10 (Feinstein 1974). Rather than intervene at a single moment (the moment of decision), the protocol submits the user to its control for an extended period, guiding a series of actions and intermediary decisions.[26] Moreover, at the moment of decision the statistical tools retract into their own statistical realm, which is not accessible to the physician. The protocol, on the other hand, never leaves its users; they can follow the protocol's path, drawn out on the pieces of paper before them.

PATIENT CARE FLOW CHART: TREATING CHRONIC LUNG DISEASE
(A summary of key steps in managing mild-moderate disease)

[Flow chart with the following boxes and connections:]

- The patient has dyspnea, cough and sputum, with diminished expiratory velocity and increased residual volume.
- Reduce lung pollution by having the patient stop smoking. Try to air condition and humidify environment.
- Start a program of bronchial hygiene. Nebulize with bronchodilator and use moisture twice daily. Make sure he's hydrated.
- Is the patient able to perform a regular program of moderate exercise? YES / NO
- Give exercises to expand the abdomen with inhalation and contract with expiration. Teach pursed lip breathing.
- Does he become dyspneic or wheeze with exercise? YES / NO
- Add an oral bronchodilator if necessary. Use a rectal aminophylline suppository or administer IV. Increase aerosol inhalations.
- Use preventive medicine: flu shots, prophylactic antibiotics in winter, aggressive therapy at first sign of infection.
- Do emotional factors or anxiety trigger an attack of dyspnea or bronchospasm? YES
- Graded exercises with low-flow oxygen, 2-4 liters per minute, may be used until the patient can perform on his own.
- Is the sputum thick and tenacious with difficulty in expectorating? YES / NO
- Try to liquify with wetting agents, mucolytics, proteolytics or iodides. First test the sputum against the agent in a test tube.
- Emotional stress affects the bronchial tree. Rx 2-4 tablespoons elixophylline, or 30 mg. phenobarbital.
- Supplement with a 5 A.M. nebulizer treatment and non-narcotic cough syrup.
- Does the patient cough excessively? YES / NO
- Does he still have trouble moving or expectorating secretions? YES
- Have the patient instructed in postural drainage exercises. Instruct a family member in chest percussion.
- Is he unable to take a deep breath? Is there a lung capacity of less than 2 liters? YES / NO
- Try a short term course of Prednisone, 40-80 mg for 2-3 days, then taper to a halt.
- Is he overdistended and not responding to other means of reversing bronchospasm? YES / NO
- Try IPPB to improve ventilation and delivery of aerosol. Proper instruction is vital since misuse may create further problems.
- Does the patient have symptoms of hypoxemia at rest? YES
- Treat with continuous oxygen, 2 liters per minute at rest, and 4 if he is exercising.
- Monitor arterial blood gases, watch for decreased respiratory activity.
- Is the patient's condition unaccountably deteriorating? YES
- Hospitalize for an indepth evaluation. Do arterial blood gas analysis and other tests for baseline information.

Figure 2.10
Feinstein (1974) gives this example of an "algorithm" for therapeutic management of chronic lung disease to illustrate the way various decisions are interwoven in a singular protocol (derived from D. F. Egan et al., "What more can you do for your chronic lung patients?" *Patient Care* 4 (1970): 18–63). Note how the protocol's course of action stretches out over several days and involves several different members of the medical staff. Copyright 1974 *Yale Journal of Biology and Medicine*. Reproduced by permission.

The Statistical Countercritique and the Development of Consensus Reports

The protocol as described above was thoroughly criticized by those who held more closely to the statistical approach. First, builders of statistical tools scorned the protocol as far too inflexible and simple a tool to be of much help in medical practice. "The algorithmic approach has limitations when one is dealing with uncertainty," Kassirer and Kopelman argued (1990). The "branching tree structure demands arbitrary cutoffs for clinical variables," thus presenting clinical situations "in a black-and-white fashion, when we all know how gray they often are" (ibid.). Protocol builders underestimate the complexity of medical practice—a complexity that can never be caught in simple, rigid recipes, where "the progression ... is always the same" (Sultan et al. 1988). Moreover, since these tools are so simple and rough, they end up being nothing but a form of additional regulation, fostering rigidity, stifling new development, and creating a "self-perpetuating bureaucracy" (Kassirer and Pauker 1978; see chapter 1 above). Statistical tools, on the other hand, are highly flexible and can in principle handle an infinitely diverse array of cases. They do not attempt to rigidly fix a whole range of actions; they intervene only when an important decision has to be made.

The second major criticism concerns the foundations of the protocols. Their design, it is argued, is often far too eclectic. Science, after all, is precise, rigorous, and universal; thus, creating ambiguous, pragmatic, and local tools is nothing to be proud of. All the local efforts at setting up protocols have only led to a plethora of subjective, conflicting rules to govern medical practice (Wennberg 1991). These types of protocols only too often rest "on a dangerous tautology" (Eddy 1990d): they conflate what physicians *should do* and what they *are doing*. They merely restate what physicians have been doing all along. This tautology can only be broken by using the universal, rational tools of decision analysis, Eddy argues. Kassirer and Kopelman (1990) agree: "... a principal difference [between algorithms and decision analysis] is that the latter is based on solid theoretical grounds, namely on probability and utility theory. If one accepts the precepts of these theories and if one accepts the numerical values for probabilities and utilities used in an individual analysis, one should, if rational, accept the outcome of such an analysis. The theoretical underpinning for algorithms is less defined and rarely made explicit."

Finally, in translating knowledge into the shorthand notation and flowchart-like structure of a protocol, Kassirer and Kopelman (ibid.) argue, "something is invariably lost." The ambiguity that results from this loss in precision facilitates misinterpretation. We "should delve much more deeply into the theory and basis of algorithm design and content before we grow more trees of this kind," these counter-critics conclude (ibid.). In their view, subordinating a statistical logic to a clinical one is putting the cart before the horse.[27]

An influential development that originated against the background of this critique[28] was the phenomenon of *consensus reports*. Attempting to enhance the "quality" of the delivery of medical care, and to promote the "timely incorporation of beneficial medical innovations into clinical practice" (Kanouse et al. 1989), the emergence of the US Consensus Development Program in 1977 cannot be disentangled from the struggle to answer the increasing public dissatisfaction with medical practice and its profession. (See also Colcock 1974 and Barclay 1978.) To select the most up-to-date means to treat and diagnose classes of patients, a group of experts on a given domain are to meet and form a consensus as to the current scientific opinion on a selected problem. This processing of scientific information, according to the proponents of this method, is necessary to fill the gaps in our scientific knowledge and to translate scientific information into practical, usable guidelines. Consensus-development programs have spread rapidly to many different countries, and protocols have been designed for the decision to do a caesarean birth, the treatment of breast cancer, and so forth.[29]

These consensus reports can be seen as an attempt to create a more "scientific" alternative to the protocols described above. The attempt is to design a universal (or at least national) protocol that is not an eclectic mixture of physicians' subjective hang-ups. The report should be maximally based on evidence attained through the clinical trial—the epitome of statistical rationality for clinical research (Jacoby 1988; Lomas et al. 1988). In addition, decision-analytic techniques can be used to find the most desirable management option (Jacoby and Pauker 1986).[30]

Nevertheless, this adaptation is not very impressive from the viewpoint of the statistical logic. The idea of the consensus report, with its attention to universality, its clinical trial foundations, and its openness to decision-analytic methods, is seen as a step forward. However, the idea of "consensus"—of filling the gaps and "adapting" scientific knowledge for practical purposes—still sends shivers down the spine of the more forthright upholders of the statistical ideal. All this tinkering just opens the floodgates of subjectivity and unscientific idiosyncrasy. "Who can say that a guideline developed by an expert panel is correct?" Eddy (1990b) asks. "Indeed," he continues, "what does a consensus of a group whose perceptions [of e.g. the effect of a therapy] might vary from 0 percent to 100 percent even mean?" The need to assemble a workable recipe, it is feared, will lead to reports based more on "compromise" than on actual consensus.[31] Moreover, the structure of the protocol has not changed. Simple, rigid recipes are simply not suitable for assessing "clinical problems that are controversial" (Kassirer and Kopelman 1990).

Finally, consensus reports tend to be much more general, much less detailed, than the protocols described above. This is no coincidence. The tight, direct, and encompassing control of the protocols for physician extenders was utterly unacceptable to physicians. Consensus reports have a very different status. In order to get the physician to cooperate, such protocols are usually rather indeterminate guidelines with a noncompulsory character. From the perspective of the statistical counter-critics, then, this is a dead-end street. Loosening the grip of a bureaucratic straitjacket will not compensate for the protocol's essential shortcomings, since this can only result in more space for subjectivity and bias.

Expert Systems: The Best of Both Worlds?

The Mind came in on the back of the Machine.
—*George Miller*[32]

Rooted in work done and alliances made during World War II, the field of Artificial Intelligence was born in the 1950s with the work of Alan Newell, Herbert Simon, and others.[33] As was briefly described in chapter 1, these authors argued that the human mind functions as an information-processing system. Rather than model the human mind to a calculating "mathematical device," as the early statistical researchers did, these authors drew on the metaphor of the programmed digital computer. (See, e.g., Baars 1986.) Simon (1979; cf. Pylyshyn 1980) expresses this position succinctly:

Like a modern digital computer's, Man's equipment for thinking is basically serial in organization. That is to say, one step in thought follows another, and solving a problem requires the execution of a large number of steps in sequence. ...There is much reason to think that the basic repertoire of processes in the two systems is quite similar. Man and computer can both recognize symbols (patterns), store symbols, copy symbols, compare symbols for identity, and output symbols. These processes seem to be the fundamental components of thinking as they are of computation.

According to these authors, the way to study human intelligence was to try to imitate human problem-solving abilities by means of a computer program.

Building on this information-processing framework, the expert system became a favorite research object of the Artificial Intelligence community in the 1970s. The turn to expert systems implied a shift in the search for a way to build programs that could behave intelligently. Earlier researchers, such as Newell and Simon (1972), had tried to look for a few general, problem-independent mechanisms underlying all human problem-solving behavior. Expert-system builders argued that these attempts of the Artificial Intelligence pioneers had not been very successful. This work "was dominated by a naive belief that a few laws of reasoning coupled with powerful computers would produce expert and superhuman performance" (Hayes-Roth et al. 1983). This was too grandiose an undertaking, expert-system builders professed. Rather, "high performance reasoning systems" can only come into being when sufficient attention is given to the *knowledge* required for expert functioning. Simultaneously, this shift to a "knowledge-based approach" also implied a shift to narrow, specialized domains of expertise. Duda and Shortliffe (1983) argue that "many human experts are distinguished by their possession of extensive knowledge about a narrow class of problems," and that "it is this very limitation that makes it feasible to provide a computer program with enough of the knowledge needed to perform those tasks effectively."

The pioneers of expert systems ventured into medicine, feeling that it was an optimal domain in which to try out their theories. The expertise in this domain—the medical knowledge of specialists—seemed optimally structured: it was circumscribed and relatively explicated; it dealt with specific facts, related in relatively clear ways (Davis et al. 1977).[34] In their turn, medical authors had not been unaffected by the prevalent tendency to see the computer as endowed with an aura of science and as a source of scientific, technological, and economic progress. The science and the practice of medicine, some of them thought, could benefit too (cf. Anon. 1961a; Ledley 1965).

A variety of expert systems (both medical and nonmedical) were developed in the early 1970s, all of them in the United States.[35] In the next subsection, I introduce MYCIN, one of the most famous of these "first-generation" systems.

An Early Promise: MYCIN
MYCIN, the dissertation project of Edward Shortliffe, was developed at Stanford University as a tool to aid physicians in the selection of antibiotics for patients with severe infectious diseases (Shortliffe et al. 1973; Buchanan and Shortliffe 1984). The required knowledge was gathered

from interviews with physicians, from review of the literature, and from files on actual cases (Yu et al. 1979). MYCIN's knowledge was contained in some 500 "procedural" (if . . . then . . .) rules, which covered the areas of meningitis (infection of the cerebrospinal membranes) and bacteremia (bacteria in the blood). Each rule embodies "a single, modular chunk of knowledge and states explicitly in the premise all necessary context" and "forms, by itself, a comprehensible statement of some piece of domain knowledge" (Davis et al. 1977). Rephrased in English by Duda and Shortliffe (1983), such a rule looks as follows:

IF:

1) the infection is meningitis, and
2) organisms were not seen in the stain of the culture, and
3) the type of infection may be bacterial, and
4) the patient has been seriously burned,

THEN:

there is suggestive evidence that *Pseudomonas aeruginosa* is one of the organisms that might be causing the infection.

The program begins by asking a series of questions of the physician who has invoked MYCIN's help (see figure 2.11). Through attempting to apply rules, the program tries to determine the type of infection and the possible organisms involved. In the case of the rule stated above, the program tries to see if the premises hold: whether the patient indeed has meningitis. The program can ask the physician directly if this is the case; if the physician does not know, it can try to infer this by applying other rules. When, finally, the premise is fulfilled (as is the case in figure 2.11 after question 14), the program checks if the second premise is fulfilled, and so on. As Duda and Shortliffe (1983) explain, MYCIN "attempts to achieve any goal by applying all the directly relevant rules. The need to establish the premises of those rules sets up new subgoals that are treated in the same way." Once MYCIN has completed its diagnostic phase, it determines an optimal combination of antibiotics to be administered to the patient (figure 2.12).

In 1979, MYCIN's medical accuracy was subjected to what Yu et al. proudly presented as "the first clinical evaluation by independent experts" ever performed of a medical expert system. MYCIN and eight specialists had to choose antibiotic treatment for ten "diagnostically challenging" paper cases. These prescriptions, subsequently, were evaluated by "prominent infectious disease specialists," who did not know the

```
Please enter information about the patient.
     Name    Age       Sex       Race
1) ** Pt219  42 YEARS  FEMALE    CAUCASIAN

2) Are there any positive cultures obtained from Pt 219?
** NO
3) Are there any pending cultures of Pt219?
** YES

        site   date       stain       method of
               collected  examined?   collection
                                      [urine,sputum only]
4) ** CSF     3-jun-76    YES         N/A

        stain of smear        morph
        [type NONE if no organism was seen on the smear]
5) ** NONE
.
.
  {more questions}
.
13) Has Pt219 recently had symptoms of persistent headache or other
abnormal neurologic symptoms (dizziness, lethargy, etc.)?
** YES
14) Has Pt 219 recently had objective evidence of abnormal neurologic signs
(nuchal rigidity, coma, seizures, etc.) documented by physician observation
or examination?
** YES

The CSF cultures will be considered to be associated with MENINGITIS.
```

Figure 2.11
An excerpt from a session with MYCIN. Source: Duda and Shortliffe 1983. Copyright 1983 *Science*. Reproduced by permission.

identity of the prescribers. Compared to the physicians, MYCIN performed well. The system did not fail to cover a treatable organism, whereas some of the physicians did. In addition, its prescriptions were deemed "acceptable" by the expert panel more often than those of the physicians (65 percent versus an average of 55.5 percent).

Notwithstanding these encouraging results, MYCIN was never used in practice. Apparently, Buchanan and Shortliffe acknowledge, it is not enough to show that a system performs well. Physicians found it irritating that they could not offer information and had to wait for the program to ask them. Moreover, MYCIN asked a relatively large number of questions

My therapy recommendations will be designed to treat for organisms that
either are very likely or, although less likely, would have a significant effect
on therapy selection if they were present. It is important to cover for the
following probable infection(s) and associated organism(s):

INFECTION-1 is MENINGITIS
+ <ITEM-1> MYCOBACTERIUM-TB [from clinical evidence only]
+ ...   COCCIDIOIDES   [from clinical evidence only]
+ ...   CRYPTOCOCCUS   [from clinical evidence only]

43) Does Pt219 have a clinically significant allergic reaction to any
antimicrobial agent?
** NO
    [Considering ETHAMBUTAL-AND-INH-AND-RIFAMPIN for use
    against INFECTION-1...]
44) Is Pt219 pregnant or breast feeding?
** NO
45) Do you have reason to suspect that Pt219 may have impaired renal
function?
** NO
46) Pt219's weight in kilograms (or <number> POUNDS):
** 51.3

[REC-1] My preferred therapy recommendation is as follows:
    Give the following in combination:
    1) ETHAMBUTAL
       Dose: 1.28 g (13.0 100mg-tablets) q24h PO for 60 days
           [calculated on basis of 25 mg/kg]
           then 770 mg (7.5 100mg-tablets) q 24h PO
           [calculated on basis of 15 mg/kg]
           [the dose should be modified in renal failure. I can adjust the dose
           only if a creatinine clearance or a serum creatinine is obtained].
       Comments: Periodic vision screening tests are recommended for
       optic neuritis.

    2) INH
       Dose: 513 mg (5.0 100mg-tablets) q24h PO
           [calculated on basis of 10 mg/kg]

    3) RIFAMPIN
       Dose: 600 mg PO q24h
       Comments: Administer dose on an empty stomach

Although I suspect that fungus may be a possible pathogen, in the absence of
definitive evidence, I will not recommend antifungal therapy at this time.
Please obtain samples for fungal, TB, and viral cultures, cytology, VDRL
(blood and CSF), coccidioides complement-fixation (blood and CSF),
cryptococcal antigen (blood and CSF), viral titers (blood and CSF). An
infectious disease consult may be advisable.

Figure 2.12
MYCIN's therapy advice. Note the additional questions and explanatory remarks.
Source: Duda and Shortliffe 1983. Copyright 1983 *Science*. Reproduced by permission.

(usually more than 50), which took time—a scarce resource for most physicians. These and other problems led the MYCIN team to abandon this project and shift their attention to other tools (Buchanan and Shortliffe 1984).

The Expert System: Encoding Clinical Reasoning

The expert system is positioned at an interesting point with respect to the tools described above. On the one hand, in agreement with Feinstein et al., the expert-system builders criticize the feasibility and desirability of statistical tools. The ideal of statistical inference, they argue, is unattainable:

> ... the failure of the use of the pure probabilistic decision making schemes lies in their voracious demand for data. ... For even a relatively small problem—e.g. 10 hypotheses and 5 binary tests—the analysis requires 63,300 conditional probabilities. (Szolovits and Pauker 1978)

Assuming, for example, that symptoms and signs of a disease are independent (i.e., that having symptom A does not alter one's probability of having symptom B) cuts this number down considerably. But although this assumption is almost always taken for granted, it is "usually false" (ibid.). Too often, the probability that a person has a given symptom *does* change with the outcomes of other tests. Many symptoms are interrelated through pathophysiological and anatomical mechanisms, Szolovits and Pauker argue; having a rash increases your chance of having fever, since "rash" and "fever" often go together in infectious conditions.[36]

Statistical tools create "artificial simplifications of the problem," Szolovits (1982) states:

> Attempts to extend these techniques to large medical domains in which multiple disorders may co-occur, temporal progressions of findings may offer important diagnostic clues, or partial effects of therapy can be used to guide further diagnostic reasoning, have not been successful. The typical language of probability and utility theory is not rich enough to discuss such issues, and its extension within the original spirit leads to untenably large decision problems.[37]

The expert-system builders, thus, join Feinstein in questioning the suitability of abstract statistical formulas for the complex, messy nature of medical practice. "Diagnosis needs to be only as precise as is required by the next decision to be taken by the doctor," Szolovits and Pauker (1978) argue. Tool builders should take the pragmatic particularities of medical practice seriously: ". . . the simple passage of time, 'creative indecision,' often provides the best diagnostic clues because [it] adds a whole new dimension to the other available information" (ibid.).

Expert-system builders join Feinstein's other major point of critique too. The statistical nature of the tools, they agree, is often directly antagonistic to what clinicians try to achieve. Statistical tools have been unsuccessful because their reasoning is devoid of clinical meaning. Expert-system builders share with the protocol makers we encountered the fundamental notion that content is primary: that all uses of decontentualized statistical theories should be subordinated to clinical experience or pathophysiological knowledge.[38] This would improve both the performance of decision tools and their acceptability to physicians. With regard to statistical tools, Davis et al. (1977) argue, it is often "not clear how each of the symptoms (or some combination of them) contributed to the conclusion. . . . The problem, of course, is that statistical methods are not good models of the actual reasoning process. . . . [They are] 'shallow,' one-step techniques, which capture little of the ongoing process actually used by expert problem solvers in the domain."

Notwithstanding this alliance with the protocol, the expert-system builders have fundamental problems with the setup of that tool. Protocols, Shortliffe (1987) sneers, "have been largely rejected by physicians as too simplistic for routine use." The rigid, simplistic structure of the protocol causes it to break down as soon as a nonroutine event occurs, so that "the difficult decisions are left to experts" (Shortliffe et al. 1979). When many factors play roles, and when uncertainty comes into the picture, "the rigidity of the flowchart makes it an inappropriate decision making instrument" (Szolovits and Pauker 1978). Like the builders of statistical tools, the designers of expert systems feel that protocols put physicians in unyielding straitjackets. Protocols deny physicians the flexibility they require when problems become difficult—when, in other words, there is a need for a decision-support tool in the first place.[39]

The expert-system builders' position is thus juxtaposed with the other tools in an intriguing way. On the one hand, expert-system designers share the view of Feinstein and others that medical practice is a nonquantitative domain, typified by its pragmatic, judgmental, clinical character. Ameliorating the practice of medicine, then, requires building on the concepts, theories, and know-how physicians themselves employ, not implementing decontentualized, "shallow statistical theories." On the other hand, expert-system builders join the proponents of statistical tools when it comes to the explicit focus on decisions. They position their tool in the flow of medical practice as the statistical-tool makers do. Protocols try to rationalize medical action by attempting to attain uniformity in diagnostic and therapeutic procedures. For expert-system builders, in contrast,

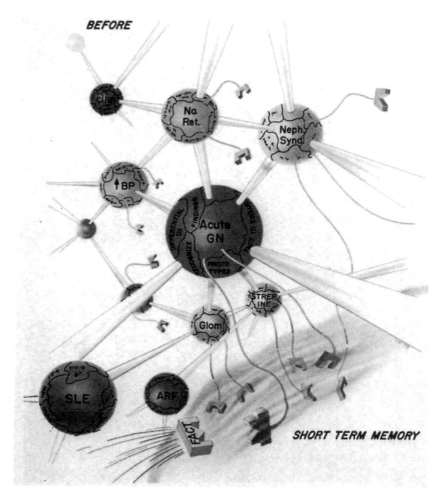

Figure 2.13
The continuous intertwinement of the construction of computer-based tools and new (descriptive and normative) models of medical practice closely follows the divergent developments of these tools. Thus, the information-processing notion that physicians organize their knowledge in cognitive "frames" or "scripts" ("packages of closely related facts") has resulted only in expert systems' being programmed with such devices (see Pauker et al. 1976). This figure describes hypothesis generation by physicians. The intertwinement of models of reasoning, of the brain, and of computers is prominent. The "frames" are portrayed as a kind of nervous cells, and words such as "short-term memory" are used to indicate parts of the program. [Original legend: Hypothesis generation. BEFORE: in the nascent condition (when there are no hypotheses in short-term memory), tentacles (daemons) from some frames in long-term memory extend into the short-term memory where each constantly searches for a matching fact.

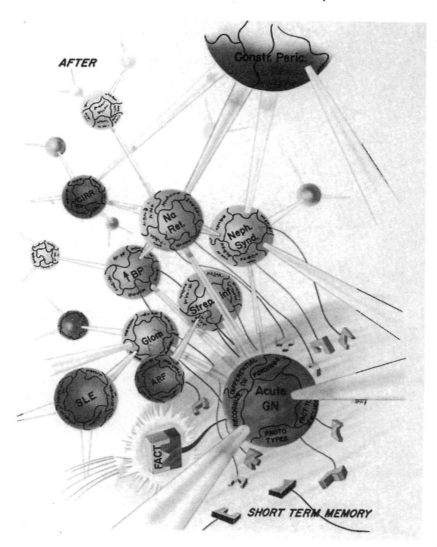

Figure 2.13 (cont.)
AFTER: the matching of fact and daemon causes the movement of the full frame (in this case, acute glomerulonephritis) into short-term memory. As a secondary effect, frames immediately adjacent to the activated frame move closer to short-term memory and are able to place additional daemons therein. Note that, to avoid complexity, the daemons on many of the frames are not shown. (Pauker et al. 1976)] Copyright 1976 Excerpta Medica Inc. Reproduced by permission.

rationality is in a fundamental way not the same as standardizing sequences of action. Rather, rationality is flexible, intelligent *reasoning*. Expert-system builders hope to optimize medical decisions without imposing restraints on more than the decision itself. Given the clinical nature of medical practice, they want to outperform the physician's cognitive decision-making capabilities—which, after all, are limited (Pauker et al. 1976; Szolovits 1982).[40]

The development of the expert system and its Artificial Intelligence predecessors, then, was inextricably tied to the emergence of the second main variant within the cognitivist discourses encountered in chapter 1. With the coming of digital computers, the physician was described as an information processor; with the further development of expert systems, the physician was said to organize his or her knowledge in "scripts" or "frames." (See figure 2.11.) Again, medical practice is redescribed in the tool's image; again, the perfect supplement to the physician is the device to which the physician is modeled in the first place.

The Statistical Countercritique Revisited: Later-Generation Systems

Since both the expert system and the protocol as depicted above embrace a clinical logic, the statistical criticisms of expert systems and of protocols ran along the same lines. As protocols, expert systems were attacked for being too superficial, eclectic, and pragmatic. Local experts' subjective opinions are just black-boxed into a computer, critics argued. No attempt is made to improve on the scientific nature of these ideas, or to make a system that is universally applicable from the outset. Modeling a physician's behavior in a computer also implies modeling a physician's biases:

There may be no assurance that information will be selected (e.g., tests ordered) in a cost-effective manner, and there may be only little assurance that gross errors will not be made. . . . In modeling intuitive judgment to avoid the problems of formal decision analysis we might be trading a headache for a case of poor vision. (Politser 1981)

Many evaluations of the "first-generation expert systems" acknowledged these criticisms. Clancey and Shortliffe (1984) state that, unavoidably, "knowledge bases are incomplete, approximate, and biased models of the world." They "always reflect the values of their designers," since they "inevitably contain judgmental knowledge relating to social costs and benefits."[41] Likewise, *ad hoc* tinkering and errors are inescapable, since in systems like MYCIN "one must include all the necessary context for a rule's application in its antecedent clauses" (Reggia and Tuhrim 1985). As a consequence, MYCIN lumps together "causal mechanisms, the taxonomic structure of the domain, and the problem solving strategies" (Buchanan and Shortliffe 1984). From a statistical point of view, of course, this tinkering mingles many matters that should have been kept separate. As Shortliffe (1991) acknowledges, MYCIN's rules "were imprecise meldings of probabilistic and utility notions." Aggressive bacteria, for example, were given more weight so as not to miss them (Buchanan and Shortliffe 1984).[42]

Several attempts have been made to design expert systems that meet this criticism. Some have tried to replace the "implicit value judgments" in their programs with explicit utility values, and the "vague guesses" with numerical probabilities.[43] This development is reminiscent of the development of consensus reports: there as well, a clinical logic was "upgraded" according to statistical standards. Yet, as was the case there, statistical critics are probably not impressed. After all, the tool's backbone still consists of clinical, judgmental, and thus "vague" and "imprecise" reasoning. If not, all the proposed advantages of the expert system, including its focus on content and its sensitivity to the pragmatics of clinical practice, would be abandoned. Adorning this basis with some well-meant gestures toward a "more fundamental" statistical rationality will not, in the view of the latter, lead very far.

Others have tried to delve deep into the "clinical logic" to find its scientific basis. The hope is to find a pure ground—a weighty, equal alternative to the clean world of quantitative inference. Duda and Shortliffe (1983) argue that what is needed is a "model of disease or clinical reasoning," and Szolovits et al. (1988) write that "humans seem to exploit several different representations of the same phenomena." Likewise, then, programs should take into account epidemiological, pathophysiological, anatomical, and clinical data and knowledge; and, as humans, they should be able to "reason from cause to effect." This can only be done through creating highly complex programs, in which for example "multiple pure hierarchies" (one for "anatomical site of involvement," one for etiology, and so forth) stand side by side and continually interact (Szolovits et al. 1988). The programming challenge, however, appears to be staggering; as of now, no system can be said to have accomplished this task.[44]

Rational Medicines

Decision-support techniques are not the unequivocal "rationalizing" tools they might appear to be. They do not jointly underwrite a homogeneous view of Rational Medicine. Rather, the "rationality" of medical practice appears to have many faces. The decision-support tools are associated with fundamentally different views of what medical practice looks like, how it should be transformed and what a rational medical practice is.

The differences between these tools do illustrate some general trends, which can be laid out in two main dimensions.[45]

First, every decision-support tool positions itself, in one way or another, in relation to the ideal of *statistical inference*. According to this view, optimal decisions can be made through the use of a mathematical formula, free from both context and specific content. The ideal-typed statistical tools form the embodiment of this notion. These tools incarnate a thoroughly statistical view of medical practice that wants to improve medical practice by replacing the physician's judgmental and erratic decisions with solid, scientific, statistical reasoning. In contrast to the statistical ideal, the builders of expert systems and protocols argue for tools which are grounded in clinical reality and which incorporate a clinical logic as opposed to a statistical one. For them, medical practice's *raison d'être* lies in its pragmatic, substantive nature—and this should be strengthened, not replaced.

The ideal of statistical inference, however, has become an epitome of Scientific Medicine. Through (for example) the institution of the clinical trial as the "Queen of Rational Therapeutics," the development of clinical science has become strongly intertwined with this ideal.[46] Thus, we see many tools that originated on the clinical side of this axis moving toward the statistical side. The development of consensus reports and the development of expert systems dealing explicitly with probabilities and utilities are attempts to bring these tools closer to the ideal of statistical inference. On the other hand, we also see movement away from this aim, as the case of decision analysis showed. Arguing the impossibility of adherence to this ideal, clinical decision analysts reintroduced "vagueness" and "subjectivity" through the notions of subjective probabilities and utilities and the content-rich structure of the tree.

The second dimension cuts across the types of tools in a different way. Both statistical tools and expert systems, I have argued, are constituent features of the cognitive discourses described in chapter 1. Both tools seek to do the work that allegedly goes on in the physician's mind when he or she makes a decision—and to do it better. In principle, then, both tools intervene only at the moment of decision. The physician gathers all the data needed, and the tools process these data in a different space—in *their* brains, as it were. The tool generates output, after which the physician takes over again.

The protocol advocated by Feinstein and others yields a different story. This tool does not primarily focus on a physician's *decision* but on the standardization of a sequence of actions. Feinstein and Weed never focused so much on individual decision making; as was elaborated in the previous chapter, their notion of scientific medical practice resides not in the mind but in the structure of medical action. In a protocol, one decision can be involved, or more, or even none—that is not crucial. What is crucial is that the protocol does not intervene at any one point in time but stretches out across a whole period. Also, since the protocol as described here does not model itself on the physician's brain (or vice versa), the "decision" is not made "elsewhere," inside some black box (a computer, or, more figuratively, some statistical formula). There is no realm where this tool momentarily retracts itself. The protocol grasps physicians by their shoulders and leads them here right, there left. Physicians may or may not understand why, but the path is laid out in front of them, on the piece of paper containing the recipe.

This second dimension is also a source of mutual quarrels. Protocol builders are accused of promoting stifling bureaucracy, while decision-focused tool builders are criticized for overlooking the wider context in which these decisions take place. And, as on the previous axis, there is movement between the positions. This movement, however, is more subtle, more elusive, and less easily witnessed when the focus is on texts describing tools. For this, we have to shift our perspective and study the construction and implementation of actual decision-support tools: the topic of the next two chapters.

A second main point of this chapter was that the diverse rationalities and images of medical practice have come into being concurrently with the development of specific tools. The statistical tools or expert systems were not called upon to fix some pre-given, long-since-recognized flaws in physicians' performances. Rather, these tools provided the metaphors for the working and failing of the physician's mind in the first place. Nor was the protocol "invented" as an answer to medical practice's problems. Feinstein's protocol was part and parcel of the standardization practices that had given medical practice its scientific footing; in bringing the protocol to the practice of medicine, notions of "medical practice" were transformed concurrently. The view of medical practice as a scientific process of distinctive, clear-cut steps is the inseparable counterpart of the notion of the protocol as an organizer of stepwise actions, as the fulfillment of medical practice's scientific character. There were no problems simply waiting for a solution: the development of the different tools was interwoven with the emergence of new rationalities and new views of medical practice. With the construction of the solutions, the specific shape of the problems was co-produced.

3
Getting a Tool to Work: Disciplining a Practice to a Formalism

Advocates of decision-support tools often sketch an image of medical work into which the tool fits smoothly. The tools, they say, do what physicians do, but better. Expert systems are supposed to capture the knowledge of the best minds available. (See, e.g., Grimm et al. 1975.) Likewise, builders of statistical tools argue that their formulas contain the best route to decision making—the route physicians would take under ideal conditions. And the protocol is depicted as a means of describing good clinical reasoning. It is said to be a vehicle through which the current standards of good medical practice can be distributed, and to merely explicate what was already implicit in the practice from which it derives. (See, e.g., Kanouse et al. 1989.) Modestly repairing the course of action taking place anyway, the proper tools (it is claimed) smoothly upgrade medical practice to its scientific status.

On the other hand, critics such as Hubert Dreyfus argue that decision tools cannot work, since what these tools attempt is not possible. In the unruly reality of everyday medical practice, critics claim, achieving meaningful action through pre-set rules and formulas is just not feasible.

In this chapter I will challenge both positions. I will demonstrate that getting a decision-support tool to function in particular medical practices involves a thorough and specific transformation of these practices. From tool builders' literature and discussions in medical journals, the scene of action now shifts to the work of designers creating an actual tool and trying to get it to operate in a specific medical workplace (or series of workplaces). Contrary to the ideal-typed views of system builders, tools do not simply slip into their predestined space within a practice. Rather, getting a decision-support tool to work and constructing a niche for it in a local medical practice involves continuous negotiations with all the elements that constitute the practice—including nurses, physicians, patients, laboratory tests, blood cells, and auscultatory sounds (sounds heard through a stethoscope).[1]

This focus throws new light on the debate between critics and advocates about the reach of decision-support tools—about which domains are formalizable and which are not. I challenge the critics' emphasis on the impossibility of these techniques, and discuss some consequences of the negotiation processes described.

To include some of the diversity in decision-support techniques, I focus primarily on three tools.[2] ACORN, a computer-based "chest pain advisor," is introduced in the following section. The second tool discussed, de Dombal's abdominal pain system (a statistical, computer-based tool), has already been introduced in the previous chapters. (Since this tool does not have a single name, I refer to it as "de Dombal's tool" or "the abdominal pain system from Leeds" or some such combination.) The last example I draw upon is a research protocol that treats patients having locally advanced breast cancer (that is, with involved local lymph nodes but no apparent distant metastases) with an experimental therapy called "peripheral stem cell transplantation."[3] These cases are compared with the cases of similar patients treated with "conventional" chemotherapy. (See figure 3.1 for an overview of this protocol.)

Because research protocols are highly detailed and widely used in medical practices, they are excellent objects for study. Also, research protocols are supposed to fulfill all the functions of a well-designed standardizing protocol (see chapter 2): they articulate actions over time and place, give detailed recommendations given specific situations, and (it is hoped) transport optimal therapeutic regimes to an array of medical practices. Their widespread use, moreover, seems to contradict the critics' repudiation of the feasibility of formal techniques.

Disciplining a Practice to a Formalism

The Story of ACORN
Acute chest pain patients constitute a well-known problem for emergency department personnel in the United Kingdom. Decisions about whether or not such a patient needs urgent admittance to a coronary care unit (CCU) must be made quickly and on limited information. CCU beds are scarce and expensive, so unnecessary admissions should be kept to a minimum. On the other hand, delaying a needed admission can be a matter of life and death. Acute myocardial infarct, dysrhythmias, and unstable angina all have a high early mortality rate. For these patients, the benefit of being treated in a CCU is greatest in the first few hours after an attack.

Delays and mistakes, however, are frequent phenomena in emergency departments. Patients are admitted unnecessarily, are erroneously discharged, or have to spend long periods of time waiting before a decision is made. There appeared, Jeremy

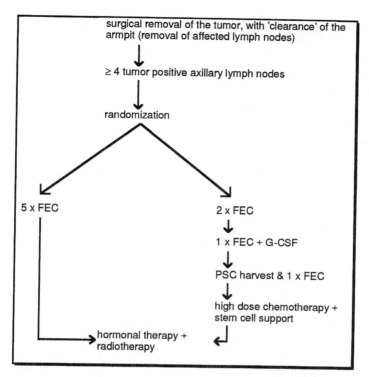

Figure 3.1
An overview of a breast cancer protocol. FEC is the abbreviation for the combination of three chemotherapeutic agents used. PSC stands for peripheral stem cell. The left arm is the "conventional" treatment; the right arm is the experimental therapy. In the latter, blood stem cells are filtered ("leukapheresis") from the patient's blood stream and deep-frozen. Subsequently, the patient is treated with massive ("ablative") chemotherapy. This would ordinarily kill the bone-marrow cells responsible for the production of the blood cells. In this treatment, however, the unfrozen blood stem cells are given back to the patient after the highly toxic chemotherapy, which is hoped to affect the tumor cells in equally deadly fashion.

Wyatt (1989) notes in retrospect, "to be a role for a decision-aid that could rapidly process relevant information about a chest pain patient, and help casualty staff to solve the urgent problem: should a patient be Admitted to the CCU OR Not?" Peter Emerson, a chest physician working in an emergency department in London, had previous experience with decision-analytic methods. He started work on a Bayesian tool (similar to the then already well-known abdominal pain system in Leeds) to help nurses make these decisions.[4]

The data items for this system had to be either examinations nurses could perform or questions the nurse could ask the patient.[5] Emerson set out to ask some of his experienced colleagues (all physicians) which items they deemed relevant. The three lists of items gathered in this way, however, were often in conflict with one another. There was widespread opinion about how significant different symptoms and signs were. "If an attempt had been made to build a system from the collected personal constructs of this group of experts," Wyatt and Emerson (1990) state, "the resulting repertory grid would have defied all known methods of analysis." Through a first rough selection as to their expected importance and ease of collection, this list was cut down to 54 items: 45 questions on the history and the characteristics of the pain and nine "simple" nursing observations about pulse rate and rhythm, temperature, blood pressure, and the general appearance of the patient. In February 1984, a questionnaire containing these items was drawn up to collect data from presenting patients complaining of chest pain.

Using this list, a database of some 400 patients was created. Analysis of this database soon made it clear that many of the physicians' insights were not reliable. Compared with statistically determined powers of discrimination, physicians often overestimated the relevance of isolated signs and symptoms. They attached much weight to the type and the duration of pain, while the statistically measured predictive value of these items for a cardiac condition requiring admission to the CCU appeared to be low. Also, many data items suggested by physicians could not be used, since the nurses were not able to elicit them consistently. Many questions seen as relevant by the doctors failed this requirement; for example, "Is the pain sharp in nature?" was found to have a repeatability of only 66 percent.[6] Finally, the large number of items ran counter to the original goal of the system: having to ask 45 questions in addition to performing nine investigations was not very helpful in reducing delays!

To deal with these problems, the designers set out to shorten the list. By eliminating items that had statistically low predictive values and/or items that could not reliably be collected, they managed to reduce the number of items to 22. With these 22 items, a Bayesian formula (see chapter 2) was devised. Figure 3.2 schematizes the setup of the system at this point.

Now, however, Emerson and his co-workers ran into another difficulty: the tool did not work. When patient data from the database were used, it appeared that this

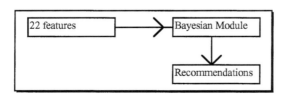

Figure 3.2
The structure of an early version of ACORN.

"simple Bayesian approach failed to produce a viable decision aid, because over 30 percent of patients fell into a middle probability band between the 'Send home' and 'Admit to CCU' thresholds" (Hart and Wyatt 1989).[7] Moreover, ACORN would sometimes suggest clearly impermissible actions, such as sending home a patient who looked to be very ill and in shock. This could happen, for instance, when the patient would further have only a few positive items, thus yielding an overall Bayesian score below the threshold of "Admit to CCU."

The designers concluded that "a system based on probabilities alone was not sufficiently accurate and that additional symbolic rules were needed to complement the Bayesian analysis" (Emerson et al. 1988).[8] This hybrid system started out just like the first version: by processing a list of (now 12) indicators with Bayes' Theorem. It decided whether the probability of the current patient's being at high cardiac risk was either low, middle, or high. Subsequently, a small expert system, containing some 200 rules, processed up to 12 further clinical features. This analysis would result in one of the following advices:

(i) admit to the CCU immediately as a case of acute ischemic heart disease. No further investigation required.

(ii) classification as a non-cardiac case not requiring an ECG (with a recommendation about whether or not to order a chest x-ray).[9]

(iii) do an ECG (of those patients of whom the computer had insufficient information to decide upon either i) or ii)).

When the last advice was given, the nurse performed an ECG, had the doctor interpret it, and went back to ACORN to enter this information. The structure of ACORN at this point is outlined in figure 3.3.

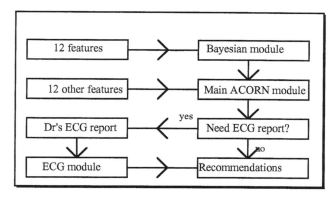

Figure 3.3
The structure of the "hybrid" ACORN (after Wyatt 1989).

This system was installed in the department and tested to see whether it would be of help to the nurses. To the disappointment of the system's designers, three new, serious problems emerged. First, ACORN performed poorly. It admitted and discharged patients wrongly far too often to be an acceptable decision aid. Second, the nurses often did not use the system immediately. When they finally entered the data, they had often already made the important decisions. Third, the nurses complained that the system was too complex and told them to do things they were not allowed to do, such as admit a patient to the coronary care unit (formally a doctor's job, they said).

Confronted with these problems, the designers made some fundamental changes to the system. First, they decided that one major reason for all the problems mentioned was that the program was still trying to accomplish too complex a task. It was trying to do too much with a relatively limited set of data. For example, ACORN tried to figure out whether pneumonia of pulmonary infarct was likely to be the cause of the chest pain—in which case it would ask for an x-ray. As Wyatt and Emerson came to realize, however, they "could not get anywhere near accurate diagnosis of non-cardiac pain with the items used" (interview, Emerson). They decided to limit themselves to the question of whether the patient could have a cardiac condition requiring acute admission to the CCU. A more general approach to the problem of acute chest pain, they felt, was not feasible.

Closely linked to this modification was the decision to stop trying to determine whether a patient needed an ECG. "It was more and more becoming practice to do an ECG on virtually anybody with chest pain," Emerson noticed (interview, ibid.). Moreover, incorporating the ECG results from the very beginning would speed up the process (which, as described above, now sometimes involved two separate sessions with the system).

To make this change feasible, yet another major problem had to be tackled. ACORN's builders found that the delay and the inaccuracy of ACORN's advice were primarily due to the fact that the ECG depended on physicians. Since the physicians' readings were often both inaccurate and late, the designers decided to eliminate this dependency. Instead, an automated ECG-interpreting machine was installed, which provided "more accurate ECG reports within seconds" (Wyatt and Emerson 1990). The nurse now had to take an ECG, answer ACORN's questions about the ECG, and subsequently answer up to 12 questions on the history, symptoms, and signs. By now, Wyatt had left the team, and Emerson had completely abandoned the statistical part of their system; it now consisted solely of symbolic rules. (A schematic outline of the new ACORN is shown in figure 3.4.)

And other changes were made. Since part of the delay was caused by the computer's being too slow and located in a separate room, ACORN's ease of use and its mobility were improved by installing the program on a fast laptop computer

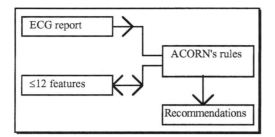

Figure 3.4
The structure of ACORN after the abandonment of its statistical part.

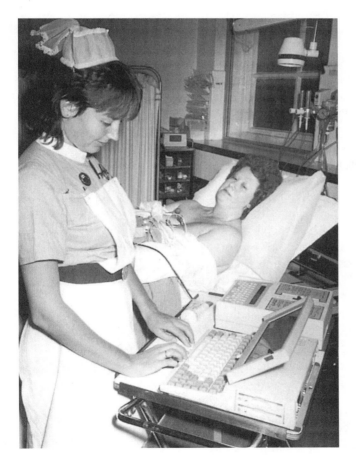

Figure 3.5
ACORN in use. Photograph by Dr. Peter Emerson. Reproduced by permission.

mounted on a trolley (figure 3.5). Finally, it was decided that the 13 advice options originally provided by ACORN "did not fit in with the nurses' 'triage' system" (Wyatt and Emerson 1990). Suggestions to perform additional tests and conclusions about possible diagnoses appeared to be too elaborate. They were reduced to three: "admit urgently," "see the doctor soon," and "wait in the queue to see the doctor" (ibid.). The designers furthermore arranged for a new policy to be formally agreed upon: when ACORN advised the nurses to do so, they could admit a patient to the CCU without the intervention of doctors.

■

The introduction of a technology into an existing practice is a process of continual negotiations. Decision-support techniques such as ACORN or the breast cancer protocol contain specific notions of what nurses, for example, are and are not allowed to do, and which patients' signs are relevant and which are not. In other words, decision-support tools have an inscribed *script* delineating who or what the relevant elements in the involved practice are and what their respective roles consist of (Akrich 1992; Akrich and Latour 1992).[10] When a technology such as ACORN or a protocol is introduced, the originally inscribed script may be challenged by any of the heterogeneous elements it affects. In the negotiations that follow, these elements, including the technology itself, can all be transformed.

At first, the nurses did not subscribe to ACORN's script: in the first field test, they often did not use the tool. The tool had to be modified so that it would be more mobile and gave less elaborate advice. Moreover, the designers arranged an official redelegation of responsibilities among physicians, ACORN, and nurses: now, when ACORN told them so, nurses no longer had to call upon physicians to admit a patient to the CCU.

The elements constituting this practice also include the patient's chests. These could also challenge ACORN's script. If ACORN would too frequently send healthy hearts to the CCU and return ischemic hearts home, it would be a failure. The chests, however, did not speak for themselves: a host of medical data spoke for them. These at first made the creation of ACORN seemingly impossible. They were too numerous, they contradicted each other, and they often were unfit for use since the nurses could not gather them consistently (or in time, in case of an ECG). Here again, the designers negotiated with these elements in order to get some version of ACORN to work.[11] To make ACORN possible, the list of items first had to be cut down. In order not to lose the physicians in the process, ACORN kept the list of medical data cited by them as its central core. Subsequently, statistical techniques were used to select the data items that spoke most clearly and uniformly—and at one point all items dealing with noncardiac diagnosis were thrown out. Finally, in order to keep the nurses' role feasible, the data items that spoke too indeterminately were thrown out, and interpretation of ECGs was delegated to a machine.

Negotiations were also part and parcel of the construction of the breast cancer protocol. In the following fragment, the setup of the protocol is discussed in an inter-institutional research meeting of oncologists. A first agreement on the desirability of this project has already been reached:

Paula,[12] who chairs the meeting, asks the group how many lymph nodes they want to settle upon. This number stands for the amount of axillary nodes in which, after the first surgical resection of the tumor, cancer cells were found. On the whole, it is assumed that the more lymph nodes are involved, the poorer the prognosis will be. Settling on ten would imply, in this case, that only patients with ten or more involved lymph nodes would be eligible. The number chosen is dependent on several different considerations. One wants to reserve such an experimental, intense treatment (with potentially lethal complications) for patients with poor prognoses. On the other hand, the poorer a patient's overall prognosis and condition, the less effect the therapy will have.

"If we set the limit to ten," one oncologist starts the discussion, "we might get too few patients. After all, the category of 'more than ten positive nodes' contains only some 10 percent of our patients. We might have too few cases for our study." "How many patients are we talking about?" somebody else intervenes. Some numbers are mentioned, but nobody is really sure. "But we can ask the national registration office," Paula says. "Although I'm actually not even sure that our pathologists do not stop counting nodes at some point. They might not even reach ten; they might get bored beforehand, and just make a rough guess. I'm not sure." Some doctors laugh, and a fourth oncologist asks: "How many patients would have more than seven nodes?" Paula shakes her head: "The registration does not register that. 'Five' or 'ten' they will know. But that will be too much; we'll be flooded with patients. We cannot do more than so many of these treatments per year." "Not if we set the eligibility age lower," the same oncologist counters. "That clears away much."

Another oncologist joins the discussion: "If you would set the number of nodes lower than, say, ten, our center would get in trouble. We've already got some local trials running for these categories; this protocol would interfere." John, the main author of the protocol, interrupts: "That's no problem. You can enter all patients you want. We will deal with the differences in entry between centers. And about the expected number: we have to realize that some 20 percent will drop out, and that we've got a control group [receiving the conventional treatment], and that we need quite a number of patients to show the small difference in survival we expect. I would argue for going for, say, four nodes. Which is also a cutoff point often used in the scientific literature—where we want to be, right? If we settle on seven, for example, we cannot compare ourselves to other studies. Finally, if we have more eligible patients, we can finish our study sooner. This is good news both for our financiers and for our publication plans—and it also supports the choice for a low number of nodes."

This episode deals solely with the number of lymph nodes to be used as cutoff point. Already in this short episode, however, it becomes clear how the script of this protocol interweaves a wide, heterogeneous range of elements. The number of nodes, a pathological-anatomical criterion, was related to the statistical power of the trial, the workload of the centers, the habits of the registration office, the position of other, local protocols, the financial situation of the whole group, their alignment with the scientific literature, the fate of individual patients, and the (not explicitly discussed) question of which patients can be asked to "suffer" this therapy. All these issues had to be aligned with one another for the protocol to become feasible; all these issues were affected by the seemingly simple choice of four versus ten nodes. The centers would get their extra share of work, and they would have to renegotiate, within their own departments, how this protocol would interact with other, already-existing protocols. Similarly, everybody had to agree to administer this treatment to the larger group of (relatively) less ill patients. If all went well, they would receive financial support, with which some centers could buy the equipment necessary to perform the peripheral stem-cell transplantation. Also, they might succeed sooner, and more convincingly, in showing that this treatment would be beneficial.

In the construction and implementation of decision tools, then, the practices within which the techniques become embedded are thoroughly transformed. In introducing their tools in the emergency department and oncology wards, the system builders we encountered here did not restrict themselves to working solely on their tool. They redesigned the wards' practices as well. They tinkered with the wards' elements in order to accommodate them to the needs of the tool.[13]

These changes are not arbitrary: decision-support techniques are *formal* tools. They operate using a collection of explicit, symbolic rules or statistical formulas, which turn input data into output. Given, among other things, the number of lymph nodes (input), the breast cancer protocol, according to its (branching) logic, delivers the judgment on eligibility

	day 21	day 28
WBC $\geq 3.0 \times 10^9/l$ and plat. $\geq 100 \times 10^9/l$	100%	100%
WBC $\geq 2.0 < 3.0 \times 10^9/l$	delay one week	75%
WBC $< 2.0 \times 10^9/l$	delay one week	off study
plat. $< 100 \times 10^9/l$	delay one week	off study

Figure 3.6
A table from a protocol indicating dose reduction or delay in the FEC regime. FEC is administered in one day, once every 21 days. As its main side effect, FEC leads to a suppression of the production of blood cells by the bone marrow. Given the number of white blood cells (WBC) and platelets (plat), the table determines what action should be taken. When, for example, at "day 21" (when a next dose should be given) the number of WBC is below 3.0×10^9 per liter (indicating poor recovery), the protocol instructs to "delay 1 week." At day 28, the white blood cells are measured again; depending on the results, the patient is given the full dose of FEC, is given a 25-percent-reduced dosage, or is sent "off study."

(output); somewhat further along its path, yet other rules determine the dosage of the chemotherapy (see figure 3.6). With ACORN, feeding in the requested patient features generates a piece of advice; in the Leeds abdominal pain system, inputting some thirty data items results in a diagnosis.

Formal tools carry some specific requirements. First, they require a well-defined set of clear-cut, elementary bits of information as input. Whether the rules are statistical or symbolic, the input items must at all times match a possible, discrete starting point contained in the rules. All the eligibility criteria of the breast cancer protocol, for example, are made exact through laboratory tests and specific checklists: the cancer should be "histologically confirmed stage IIA, IIB, or IIIA adenocarcinoma of the breast," the performance status should be "ECOG-ZUBROD 0 or 1" (appendixes of the protocol tell you what these codes mean), the "creatinine clearance" should be "≥ 60 ml/min" (a measure for renal function), and so on. There is no room for "The kidneys look all right" or "This patient is between stage IIA and stage IIB"; the protocol has no rules to deal with such statements. Similarly, the current version of ACORN asks "What is the site of the pain?" It then draws fourteen little puppets on the screen, each with a different shaded area over its chest. The program will not proceed until the nurse has chosen one of these puppets. ACORN also imposes a definite set of input items and a definite sequence in which they are to be obtained. The ECG comes first, for example; without an electrocardiogram, ACORN does nothing. Second, the output of formal systems is predefined within the rules. Generally, a system contains a circumscribed set of pre-fixed statements, from which it selects one or several.[14] ACORN's final advice always consisted of one of at most thirteen possibilities. And often the protocol merely says either "yes" or "no"; sometimes, specific branches might make intermediate options possible (as when, in the case of a reduced white-blood-cell count, the FEC dosage is reduced by 25 percent—see figure 3.6). A seemingly innocent consequence of this feature is that, at all times, all possible output must be feasible in the practice in which the tool functions. If, for example, one of the output statements requires the performance of a laboratory test that is not available in that particular setting, the decision tool is useless. Similarly, staff members need to have the time and skills to follow the tool through. Users have to be persuaded to act upon the advice—and to do so in the way intended by the formal tool (cf. Collins 1990). The hybrid version of ACORN, for example, sometimes offered the advice to obtain an ECG. When the users of the system always already do that, however, the whole purpose of this advice becomes meaningless.

A formal system thus requires disciplined users to enter well-defined input, and expects the practice to be able to live up to pre-set output statements. In addition to these demands, a formal tool also requires a certain stability in the relations between the input data items and the content of the output statements. A data item like "the location of the pain" is of use to ACORN or de Dombal's tool only if it can be linked, through one or several (mathematical or symbolic) rules, to one of the output statements the system contains. If the location of the pain has no relation whatsoever with the cause of abdominal pain, it is obviously senseless to include it in the Leeds data item list.

So, a data item is useless if the existence of a relation is dependent on a host of contingent factors. For example, the breast cancer protocol sets the time for the leukopheresis (the "harvest" of blood stem cells) at the moment that "CD34-positive [blood] cells are detectable in the peripheral blood." (CD34 stands for a group of immunological markers which, according to the protocol's authors, constitute a standardized way of checking the presence of blood stem cells: when a certain level of CD34-positive cells show up in the blood, the chances of gathering sufficient stem cells are high.[15]) The amount of CD34-positive cells, however, can be linked to leukopheresis in the breast cancer protocol only if all institutions measure these cells with the same techniques, and if the different immunological markers used indeed function identically—something which was questioned by some of the oncologists. If not, centers might be harvesting stem cells at different times—which might have unanticipated effects on the success of the harvest and the outcomes of the treatment. In such a situation, an item like "CD34-positive cells" cannot be used by a formal system since it does not behave predictably enough. Similarly, if one is building a system like ACORN, the question whether the patient can be admitted to the CCU cannot depend on too many circumstances. Whether the CCU is full or is expected to be full, whether the patient wants to be admitted, or whether the patient lives near another hospital where he or she would prefer to be sent to cannot be too important. If many such contingent considerations played a role, there would again be no way to link the input data items with the possible advices given as output.[16]

All in all, thus, the tools require many of the diverse elements constituting the medical practices to behave in a uniform, stable, and predictable way. A broad range of elements are affected: ensuring the proper execution of the CD34 test is just as much a matter of getting the laboratory staff to act in similar ways as a matter of ensuring that different immunological markers behave identically. Similarly, administering the FEC cycles requires both the patients and the hospital organization to adhere to this "rigorous three-weeks schedule"—and to be ready for a one-week delay when the blood cell count so indicates. It is not only the "nature of the clinical problem" that has to be well defined, as computer-based system builders often argue when justifying the choice of a specific "domain" of interest.[17] It is not just a matter of finding a group of diseases with clear-cut symptoms and explicit criteria for deciding on diagnosis and therapy. It is *a whole, hybrid practice*—including nurses, physicians, data items, and organizational routines—that must be made sufficiently docile.

Moreover, whether a practice is disciplined enough for the tool to work is not a pre-fixed, given fact. In the active "redescription" of the oncology and emergency practices, many heterogeneous elements needed to be *transformed* to make their behavior definite, uniform, and predictable enough for their protocol to work. A sufficiently disciplined practice is an *actively achieved accomplishment:* pathologists have to be instructed to count precisely, CD34 has to be measured in a similar way in all centers, and the medical personnel involved has to be taught to meticulously measure and document "side effects." In the case of ACORN, the ECG-interpreting machine and the system's answer options "digitized" the input, and the department's organization was restructured to make ACORN's output feasible.[18]

Similarly, to ensure suitable input, the group from Leeds had to train physicians so that they would all mean the same thing when stating, for example, that the pain was "severe." Without adequate training, the observer variation was far too large for any decision tool to work with. When three untrained physician observers witnessed one clinician interviewing a patient, they "were unable to agree in 20.4 percent of circumstances as to whether or not a particular question was asked. [Moreover, they] were unable to agree in 16.4 percent of instances as to whether the patient's answer was positive or negative." (Gill et al. 1973) As de Dombal's team states, it took new staff "some six to eight weeks during which the personnel familiarized itself with the terminology," and during this period "the system performed much poorer" (de Dombal and Gremy 1976).[19]

In the implementation of a decision-support technique, then, the involved practice is *disciplined to this formalism*. The networks of elements constituting the medical practices involved have to be made sufficiently "tight" for the tools to function (cf. Callon 1991). Instead of delimiting what decision-support tools can or cannot do, thus, I argue that we have to focus on how domains are *made* "formalizable." Getting a tool to work is neither impossible nor a simple actualization of an already-perfect match. The limits here are not pre-set. We are, rather, dealing with a moving frontier: a place of struggle to fulfill the prerequisites of formal tools.

Building Simple, Robust Worlds

Writing a good protocol is not so difficult. What is hard is getting and keeping it in place.
—Dr. Bert Howard (oncologist), *in an interview*

In disciplining practices to decision-support techniques, several recurring patterns can be traced. In this section I point to three related ways through which attempts are made to render elements uniform, stable, and predictable: reinforcing bureaucratic hierarchies, materializing the tool's demands, and shifting decision power to the most uniform and predictable elements. Together, these patterns emphasize the accomplished and heterogeneous work of getting a tool to function in an actual workplace.[20]

Reinforcing Bureaucratic Hierarchies

A first recurring pattern in the disciplining of practices to a formalism is the installation or reinforcement of specific bureaucratic hierarchies (cf. Horstman forthcoming). In the case of the breast cancer protocol, eligible patients have to be registered at a national "trial office." Here the randomization takes place, and the name, age, diagnosis, and number of positive lymph nodes are registered. Also, it is checked whether "informed consent is obtained." The patient's therapy is thus determined at the national level; the physicians involved do not even get to do the randomization. Moreover, registration ensures that, from that moment on, the patient is "entered": the physician will now have to explicitly justify every nonprescribed action. For every "protocol-violation" the study coordinator will have to be contacted.

Hierarchical relations can also be installed or reinforced *within* institutions (cf. Kling 1991). The following fragment from one of the centers participating in the breast cancer study illustrates how supervising relations among physicians are often enforced to ensure compliance:

> In a discussion between oncologists and residents, one of the residents asks whether they should not be given the freedom to modify chemotherapeutic dosages when side effects occur. This resident works at the oncological outpatient clinic, where she is responsible for the administration of these medications. One of the oncologists disagrees: "I feel that we cannot give that responsibility to you. If we write a prescription, you have to be able to follow that blindly. Often some protocol is involved, which you might not be aware of; in these cases different rules apply for whether or not you can continue treatment in the light of side effects. So, I would say that you just call us when you're not sure. Don't go changing things on your own." (participant observation notes, 1991)

Finally, it is only through this rigorously hierarchical setup that many decision-support tools are feasible at all. Both in the case of Greenfield's upper-respiratory-tract-infection protocol and in the case of ACORN, supervising physicians have to be constantly available as a "backup." They have to deal with all patients the health assistants or the nurses cannot deal with by themselves, and intervene whenever contingencies not foreseen by the tool occur. So, while here health assistants and nurses get to do tasks they were not allowed before, this is only possible through a thoroughly hierarchical anchoring of these newly gained responsibilities.

Materializing a Decision Tool's Demands

Embedding a decision tool's exigencies in material arrangements as instruments and other artifacts is a subsequent recurring, often effective means of ensuring compliance. Materializing a tool's demands prestructures the medical personnel's work environment so that the decision technique becomes an unavoidable (and often unnoticed) part of daily practice (cf. Fujimura 1988; Suchman 1993b). By requiring the use of specific forms, for instance, medical personnel can be directly guided in the taking of a history or the sequencing of a therapy. De Dombal's team introduced a specific form in which "the patient's case history could be to some extent 'formalized'" (Horrocks et al. 1972; see figures 2.1 and 2.8 above). Through the use of this form, they forced investigating physicians to enter their findings in the required "structured and well-defined" way (de Dombal 1990). This "structured way" also implied another requirement secured through the form: all data items had to be entered at the same time. Physicians could not fool around with this demand: they had to hand over the form to a research assistant, who encoded the form and sent it to the main computer. Finally, de Dombal argued, to ensure ongoing commitment it should be made impossible *not* to use the form:

The trick is to get the form used as a permanent part of the case record. That is absolutely crucial. If the young doctor [the emergency ward resident] has to fill out that form and then write it out again in the case record, they won't do it. (interview, July 18, 1993)

Forms are only one way of materializing a tool's demands. In the next fragment, a different material arrangement ensuring the strict performance of a protocol is already in place:

When nurses change the "Hickmann catheter" [a catheter inserted in the subclavian artery, just below the collar bone, through which chemotherapy and bloodcells can be easily administered], they always use what they call the "Hickmann set." Usage of this set is prescribed by the protocol on Hickmann catheter-replacement. The "set" is a sterile, pre-assembled package containing the material required for changing the catheter according to the protocol, such as bandages, small trays for the disinfectant, tweezers, and so forth. (participant observation notes, 1991)

Similarly, the tools' output possibilities can be materialized. In research protocols, for example, medications can be pre-packaged and centrally distributed. In this way, participating institutions obtain the (expensive) drugs free of charge, in a form adjusted to the protocol at hand.

Materializing a tool's demand is often intertwined with the reinforcement of hierarchies: after all, hierarchies can be (and often are) materialized. Forms requiring signatures from superiors, pre-packaged chemotherapy that can be modified only through consultation with the study coordinator, and other materializations embed and sustain hierarchical relations. This intertwining is forcefully illustrated by the protocol for "physician-assistants" designed by Harold Sox Jr. et al. (1973), where the assistants' actions were automatically monitored. To check compliance, the checklists filled in by the physician-assistant were screened by a computer for omissions and errors in following the protocol's branching logic (Tompkins et al. 1973).

Reshuffling Spokesmanship: Shifting Decision Power among the Elements
A third recurring pattern in the disciplining of a practice to a formalism concerns the decision power encoded in the tools: the input items the formalism weighs heavily in choosing a branch in the protocol, selecting advice, or reaching a diagnostic statement. Here we see that tool builders have a preference for "trustworthy" elements: the elements that, from the perspective of the tool builder, exhibit the most predictable and unequivocal behavior. "Trustworthy" does *not* imply "better" or "more true," but points at the tool builders' tendency to delegate decision power to the signs, symptoms, and tests that best, or most easily, fit the formal tool's prerequisites. As a result, spokesmanship is often redelegated from staff and patients to laboratory tests or machines. Rather than letting physicians decide when to harvest peripheral stem cells in the breast cancer protocol, CD34-positive cells settle the matter. The protocol builders deemed the criteria used by individual physicians too idiosyncratic, and they saw the number of CD34-positive cells as a test that would yield identical results in all centers. Rather than allow "clinical judgment" to have its way, also, the number of lymph nodes and other quantitative, laboratory-derived values determine eligibility for this protocol. Similarly, rather than have physicians interpret ECGs for ACORN, the designers opted for an automated ECG interpreter, thus ensuring fast and unequivocal results. When certain "non-human" elements behave predictably and uniformly enough, decision power is often taken out of the hands of the physicians and nurses (frequently deemed hard to discipline) and handed over to them. More often than not, in search for the most unequivocal and constant elements available, a laboratory test provides the more uniform, "digitized," and predictable behavior.

Likewise, patients' abilities to control the specific course of events are often limited. Patients can always refuse cooperation with a decision-support system or step out of a protocol—but that is the only active role they can take. Aside from that, they have little room for influence.[21] In the breast cancer protocol, the patient signs a form indicating that she consents to participate in the trial. She is told that she has "the right to withdraw cooperation at any moment." But in the tight network required for the protocol to work, there is no room for additional desires such as a somewhat less intense second course of chemotherapy. If the laboratory tests do not register side effects, the second course will be the same as the first; if they do, the dose modifications are already precisely prescribed. In ACORN and in de Dombal's tool, similarly, patients can only answer pre-set questions about the location and duration of the pain. The pre-set output statements cannot deal with patients' idiosyncratic desires or needs: they are geared toward a generic patient.

In addition to a shift away from the voices of patients and medical personnel and toward cells and chemical reactions, there is a selection of the most trustworthy element among the physical signs and laboratory tests. Laboratory tests and physical signs are not more trustworthy *per se*: as many studies have shown, their robustness is itself the result of much work.[22] The CD34 test is debated among oncologists exactly because its trustworthiness is debatable. Since the centers might use different techniques to measure these cells, an oncologist remarked during the research-protocol meeting, "center [A] might be looking at something quite different than [B]." Similarly, data items should speak in a clear voice: a large part of ACORN's story was concerned with the selection of the data items that would speak for the heart's condition most unequivocally and constantly. The data elements selected were those that both spoke clearly enough for the nurses to gather them and were elected by statistical techniques as "truly representative." Auscultatory sounds, the classic voice of the heart from the physician's perspective, were not included since nurses could not elicit them. In an interview on the development of the Leeds acute abdominal pain tool, de Dombal recounted a similar story:

> First we created a long list with items mentioned in the literature. Then we got rid of those items the majority of our clinical colleagues wouldn't do, or where they could not agree on the method of elicitation. The reproducibility of the item is important: we have thrown out typifications of the pain as "boring," "burning," "gnawing," "stabbing." They haven't gone because people don't use them—they've gone because people can't say what they are. We do observer variation studies on these things, and we ask people to define them. And if people can't agree what the definitions are, then we kick the items out. The observer variation on burning, stabbing and boring pain is about 50 percent—they're useless. Another example which fell off was back pain with straight leg raising: an often mentioned sign. But nobody agrees on what they are talking about. What should the result of the test be. A figure? The angle the leg makes with the table? But when should the angle be measured? When the pain starts or when it becomes bad? What angle to measure exactly? How to lift the leg? Or should the patient lift the leg himself? We could not get a group of rheumatologists, orthopedic surgeons, and general practitioners to agree about what they should call "straight leg raising," so we abandoned that. (interview, July 18, 1993; edited somewhat to increase readability)

Through a plethora of means, thus, practices are disciplined. Whether through reinforcing bureaucratic hierarchies, through materializing the tool's demands, or through delegating decision power to stable and uniform elements, potential obstinacy or unpredictability is averted.

It is important to avoid a specific misunderstanding. I am not claiming that the disciplining of a practice is a purely local, consciously planned endeavor of the tool builders and/or some supporters. The process of getting a decision tool to work and of building a niche in a local network cannot be understood only by focusing on those who attempt to implement the tool. Of crucial importance is the way these local networks fit into other, intermeshed practices. Oncology practices, partially through their historical tie with research protocols, are already heavily pre-structured in ways congenial to new protocols. Medical personnel are used to these intervening instruments, data items or criteria are often taken over from earlier protocols, and central registration agencies are already in existence (as was alluded to in the excerpt). Moreover, research protocols often thrive on the existence of large institutions that aid in the proper execution of protocols, financially reward compliance, distribute centralized equipment, and so forth.[23] Similarly, practices gradually become more amenable (more "pre-disciplined" in the required fashion) to computer-based decision tools through developments such as the expansion of administrative bureaucracy, the efforts at standardization, the increasing use of uniformly "packaged" technologies, and the increasing computerization of laboratories.[24] These phenomena, including the crucial, active role of other participants in the involved practices, are discussed later in this book. In the following section, I look again at the transformations occurring in the "rationalized" practices in order to discuss some consequences of the recurrent disciplining patterns I have pointed at.

Changing Practices: The Specific Exigencies of a Formal Tool

In the disciplining of a practice to a formalism, we have seen, there is a tendency for individual nurses, physicians, and patients to lose direct influence on the course of events. Hierarchies are reinforced and material settings are rearranged so as to ensure compliance with a decision tool's demands. Physicians have to hand over the judgment whether a patient should enter a protocol to an interplay of laboratory values and lymph node counts. Similarly, bringing the patient's own voice back to either "yes" or "no" brings about the unequivocality needed for formal tools to work. In the new network, these actors are repositioned: inevitably, the requirement for stable and predictable elements predisposes the taming or even the silencing of these potential sources of contingency (cf. Star 1989a).

These changes point to a fundamental challenge to the views of the tools' advocates. Decision-support techniques are not the inert media they are proclaimed to be. They are not simply "carriers" of optimal knowledge, mirroring the reflection of a good practice to wherever it is needed. Rather, they transform this reflection by merging it with the exigencies deriving from its own formal structure. In these transformations, *medical criteria are specifically altered.*

The age limit in the breast cancer protocol, for example, was set at 55—a straightforward decision criterion for inclusion or exclusion. In a nonprotocolized situation, such a strict cutoff point would not have been regarded as meaningful. As one oncologist remarked, it might have been better to differentiate between pre- and post-menopausal women. "Biologically spoken, it's a different disease," he suggested. Determining the "menopausal status" (through measurements of the levels of sex hormones), however, is "tricky," as another oncologist stated. It is time-consuming, it is not always reliable, and it has none of the simplicity and clearness of simply setting an age limit. It would thus endanger the smoothness and tightness so crucial for the optimal functioning of the protocol.

Similarly, counting the number of lymph nodes is an easily performed test that will yield similar results in different institutions—if the pathologists do their counting properly. As a criterion for suitability for treatment, however, it is a rather rough and blunt means of distinguishing between patients. In medical practices most such tests are rough and blunt: there are no simple ways to determine precisely how far cancer cells have spread. However, when not working according to a protocol, physicians will often try to construct some image of the extent of the spread through physical examination (can small tumors be felt elsewhere?), clues in the history (pain may indicate spread), various imaging techniques (are the bones affected?), and so forth. In this way, some more specific idea may be generated of whether, and, if so, to what extent, cancer cells have disseminated to distant organs. This plethora of methods, however, is much less clear-cut, much harder to pre-define, and much more difficult to standardize than counting the number of nodes.

Finally, deciding on admittance to the CCU through the analysis of an ECG and up to twelve clinical and historical features puts much weight on a highly delimited number of data items. Auscultatory data and x-rays are omitted, as are questions about the patient's' home life. Again, these additional data can and often will be used by a clinician to determine whether acute admission is required—but their richness is too indeterminate and unstructured to be tamed. To function, ACORN ignores such sources of information and leans heavily on those it does take into account. So, physicians often reckon with the social history of a patient in their decision to admit a patient or not. If a patient is panicky and has an unstable home situation, they may decide to admit him even if his condition is not alarming. They might even quote a "social indication" as the reason for admittance (see chapter 5). ACORN, however, has no room for input information such as this. In its own way, ACORN takes the patient's history very seriously: for example, it tends to admit a patient who has had a myocardial infarction or is known to suffer from angina pectoris. As Emerson remarked, "if they have had it before, it takes a lot to get the computer on a different track" (interview, July 21, 1993). Whereas physicians might say that a certain person is merely anxious because of a previous heart attack and worries too much, ACORN would admit him.

Again, these examples do not show that these tools are wrong or stupid. I am not arguing here that they miss a crucial point. Rather, these examples illustrate how formal tools cannot be conceived as inert carriers of some "good medical practice." Delegating tasks to a formal tool *transforms* the nature of those tasks. The introduction of a decision-support tool generates a propensity to refocus medical criteria on the elements that behave in predictable and easily traceable ways. Formal tools contain a predisposition to build *simple, robust worlds*, without too many interdependencies or weak spots where contingencies can leak back in.[25] In doing so, in selecting the measurements and indications that best fit its prerequisites, the breast cancer protocol *redefines* what eligibility for bone-marrow transplantation treatment denotes—and, thus, what "potentially curable disseminated breast cancer" is.

Settling on a minimum of four lymph nodes as inclusion criterion selects a different group of patients than inclusion criteria based on extensive testing for dissemination would: the lymph nodes have become the salient sign determining the patient's fate. A patient with three positive lymph nodes but a large, fixed lump in the breast, for example, might be deemed a good candidate for this bone-marrow treatment by some physicians, but the protocol would rule her out.

Likewise, opting for the CD34 test above the physician's judgment transforms what it means to be suitable for stem cell harvest: the two methods will yield different results in many patients. The age limit, also,

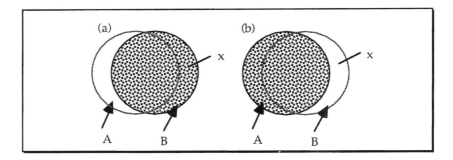

Figure 3.7
This diagram (after Wulff 1981) illustrates the subtle shift in the medical criteria for "treatable breast cancer" or "chest pain requiring admission to CCU," (a) before and (b) after implementation of a decision-support tool. The shaded areas indicate the group of patients to whom the diagnostic label is applied; the shift from A to B indicates the subtle, often unnoticed change in the use of this diagnostic label. Patient x would be treated (or admitted) in (a), but not in (b)—an effect due entirely to the specific characteristics of the tools in question.

carves out significantly different cutoff lines than a measurement of menopausal status: different criteria are invoked, which will create different groups of included and excluded patients, and rewrite the significance of particular signs and symptoms.

Finally, focusing on the ECG and some additional clinical and historical features transforms the indications for admittance to the CCU. Social considerations can no longer play a role, but a history of angina or myocardial infarction gains importance. This again rewrites the significance of these signs and symptoms, and again carves out a different dividing line between patients who are admitted and patients who are not.[26] (See figure 3.7 for a graphical representation of this argument.)

How should these redrawings be valued? In her study of decision making in neonatal intensive care units, Renée Anspach (1993) comments critically on the development of decision-support tools that utilize statistical formulas to predict an infant's prognosis:

[These developments] may take place at a certain price and may result in the loss of certain potentially valuable sources of information. For only the measurable features of infants' conditions are [usable for the formulas].... Precisely because many of the interactive cues nurses note are rarely entered into patients' charts systematically, they cannot be validated as prognostic signs.

"Subjective" signs, she argues, will then lose out against an ever increasing emphasis on technological, "hard" data.

The tendency to build simple and robust worlds, nevertheless, is not necessarily equivalent to relying solely on technological data items.[27] In the case of the breast cancer protocol, the age limit of 55 was preferred as a cutoff point over the measurement of hormones: here, a "technological" source of information was *not* used. Similarly, de Dombal's tool uses only data obtained by means of a questionnaire and a simple physical examination; his team has refrained from using any of the additional technologies (e.g., laboratory tests or x-rays) that are available. A final example: The designers of a research protocol I studied were not able to define precisely how to use "sufficient pulmonary function" (as defined by a set of technological, pulmonary tests) as an entry criterion without losing either too many or too few potential patients. As a way out, they decided to leave admittance to the trial to "the physicians' discretion." Sometimes, then, technologies are more unruly than humans; sometimes, "technological" data are less disciplined (or more costly to control) than "subjective" information.

More fundamentally, opposing "soft" to "hard" data and "subjective" to "technological" information draws a dichotomy that we should instead attempt to overturn.[28] These are not two wholly different types of data; rather, the difference is a gradual one. In fact, the argument that formal tools cannot deal with "soft" data is nothing but a tautology, since *what we call "hard" data are simply those data whose production already has been disciplined.* This disciplining, we have seen, is a highly heterogeneous affair; "hard" data are robust only because so much work has been put into stabilizing them. There is no intrinsic "softness" or "hardness" to data: what *is* measurable (and by what or whom) is the *outcome* of the negotiation processes involved in the construction and implementation of new diagnostic procedures, for example—or new decision-support techniques.

The disciplining of a practice to a formalism, moreover, is not a process to be condemned in and of itself. The often-drawn, generalizing conclusion that formal tools are "dehumanizing" or "objectifying" misses the point. Through disciplining the nurses to ACORN, their responsibilities may be enhanced; through the implementation of this tool, physicians may be relieved of some duties they did not really cherish. Through their compliance to the breast cancer protocol, residents get to handle promising new drugs they would otherwise not even have heard of, and some patients might willingly give up some right of voice in exchange for a potential straw of hope. Decision tools open up some worlds and close others; disciplining nurses, for example, might relieve them of a prior form of discipline (in the case of ACORN, for example, having to find a doctor for each and every admittance decision). On the other hand, situations may come up where Anspach's criticism should be taken at heart: when the disciplining of nurses, for example, is only intended to make someone else's work easier (see chapter 6 below). Especially when the scope and depth of the transformations are taken into account, global judgments as to which worlds are preferable are hard to make.

Finally, the disciplining of a practice to a formalism is never complete. The technologies do not simply transform medical practice to their liking: in their construction and use, they are themselves transformed. The latter transformations are the topic of the next chapter.

4

Of Nodes, Nurses, and Negotiations: The Localization of a Tool

In light of these concerns this guideline seems crazy.
—David Eddy (1992)

Some years ago, David Eddy worked on the development of a guideline on the use of contrast agents (used in radiology to visualize e.g. the stomach) for a large group-practice health-maintenance organization. Decision-analytic techniques were central to his efforts. The issue was whether the benefits of using an expensive type of contrast agent (which caused fewer adverse reactions) would outweigh the money saved by using the cheaper type. In the course of creating the guideline, several problems were encountered. It appeared difficult to figure out what the exact adverse reaction rates were. Eddy and his colleagues had already decided to divide the group of patients into two: those who had a high risk of adverse reactions[1] and those with a low risk. Now, "everyone knew that the risks were higher with ionic agents,[2] but there was uncertainty about how much higher." Analyzing the evidence in the literature appeared problematic: as is always the case, "virtually every study has something wrong with it." Study designs were imperfect, or did not match the group practice's local conditions. Moreover, hardly any study divided patients into low-risk and high-risk groups, requiring much decisions about how to interpret the data.

In addition, estimating the money saved was difficult: "To estimate the cost of a moderate reaction such as nausea and vomiting, we would have to conduct time-motion studies to determine how long an episode of vomiting ties up a radiological suite and the cost of janitorial service."

Comparing costs and benefits, subsequently, proved forbidding. Taking the decision-analytic premises seriously, Eddy set out to question members of the health-maintenance organizations about their preferences. Questions like "Would you be willing to pay more for an x-ray in order to reduce the chance of side effects?" were prepared. Slightly differently phrased questions, however, yielded completely inconsistent answers.[3] Members would say "no" to the question mentioned, but they would simultaneously be willing to pay slightly increased premiums to ensure *getting the expensive agent whenever they would need it in the future. In addition, members would simply misunderstand the questions asked—however much effort was spent in making them clear-cut.*

After much time and deliberation, Eddy and his colleagues ended up with a guideline recommending use of the cheap agent for low-risk patients. Reflecting on the process, Eddy stated that this application of decision-analytic techniques "should have been about as simple as any application . . . ever could be." It was methodologically simple and politically desirable. Still, they "agonized over this guideline for more than a year." Being explicit about all benefits and costs—making the invisible visible—appeared incredibly complex and often impossible. In addition, the radiologists remained highly reluctant to not use the expensive agents, no matter what Eddy's calculations said. They were the ones who saw the patients face-to-face; they were the ones who felt they were "withholding optimal care" from their patients. Actually, Eddy's calculations demonstrated that, even for high-risk patients, using the expensive agent as a standard would not yield benefits justifying its costs. However, Eddy stated, "we had pushed rationalism about as far as we could on this problem. Words like 'high risk' have a psychological impact far beyond their actual numbers . . . and we were approaching burnout." (derived from Eddy 1992)

■

In chapter 3, I outlined how in the processes of construction and implementation a practice is transformed after the image of a decision tool. In and through the continuous negotiations, however, the decision-support technique is transformed too: in getting a tool to work, it is inescapably *localized*. Inevitably, tools that function in a practice have had to give up much of the original, ideal-typed ideals about the power, range, and/or transferability of the tool. Many tools appear to work only in one specific medical practice instead of in a broad range of practices; other tools end up confining themselves to a small part of the spectrum of medical problems they may have wanted to address. Localization, I argue, occurs in one or more of three (often intertwined) ways: in *space*, in *scope*, and in *rationale*. The phrase "localization of a tool" points to the double meaning of the term "localization": the tool's projected universality diminishes, and the tool becomes more and more particular. Reducing localization implies increasing the disciplining of the involved practices to the formalism—but in the unruly world of medical practice this is more often than not wholly infeasible. Localization, thus, is a widespread and inescapable phenomenon, continually thwarting attempts to spread the use of these tools over broader terrains.

Localization: "One Starts with Great Expectations"[4]

Every protocol is a political compromise.
—*interview, Thomas Bear (oncologist)*

Localization in space is a process that begins with the conception of the technique and continues well after its introduction into practice. The script of the final ACORN can be read as reflecting the continual need to give in to local needs in order to get ACORN incorporated in the emergency department. ACORN uses rules and data items elicited from local experts (and then only those items elicitable by local nurses, with locally available techniques, and speaking for local hearts[5]), its advice is adjusted to local customs, and so forth. Likewise, many protocols are developed by groups of local physicians, resulting in protocols reflecting local ideas about "what is best," local organizational configurations, local skills, and local patient characteristics. Both Eddy's guideline and the upper-respiratory-tract-infection protocol of Greenfield et al. (1974) are such tools. As Greenfield et al. themselves assert, their protocol "is derived from local experience and [local—MB] peer consensus." When it would be used in different contexts, it would need "modification both to suit the needs of the individual physician and to accommodate changes in causal factors, local conditions, and therapeutic and laboratory advances" (ibid.; see also Komaroff et al. 1973).

A system can become so thoroughly imbued with specifically local idiosyncrasies that it can work only at one site. In fact, this is what has happened to most decision tools. Only very few expert systems or computer-based statistical tools have been able to deal with "the unresolved question of transferability of a successful system from its initial development site to other geographically distant locations" (Reggia and Tuhrim 1985). The large majority of computer-based decision tools are in operation (if at all) in just one location. Similarly, in overviews of the current state of the art of protocol development, authors complain that the explosion of local initiatives has resulted in a chaotic multitude of often incompatible local protocols:

> Groups currently developing guidelines vary widely in the goals, methods, formats, and degrees of precision of their guidelines. Current guidelines from different groups contain important clinical differences on such common topics as screening for breast cancer. (Pearson 1992; see also Audet et al. 1990)

However, some tools have explicitly attempted to overcome this confinement to individual sites. De Dombal's abdominal pain tool is one of the very few computer-based tools to have (partially) succeeded in doing so. Also, inter-institutional protocols have been construed; research protocols are prominent examples. National institutions, finally, have developed "standards" or "consensus reports" (see chapter 2 above) to be used by every physician dealing with the specific clinical problem addressed. Focusing on such tools sheds light on what these efforts amount to.

One phenomenon that de Dombal and his co-workers continually ran into is that "databases don't travel" (interview, de Dombal). Their system calculates the probability of a given patient's having disease D by drawing upon a database listing past patients' symptoms and diagnoses. Every time the system was implemented in a different hospital, de Dombal stated, it was necessary to check whether the computer still performed as well as it did in Leeds—and usually it did not. When the system was tested in a Swedish hospital, for example, the computer's accuracy dropped from 80 percent to 50 percent. There are many reasons for this phenomenon. For one, it often happened that a different type of patient was encountered. Sometimes the system was located in an emergency department, sometimes in a surgery ward. One difference between these locations is that patients can just walk into an emergency department, whereas all the patients in a surgery ward have been referred by physicians. This preselection thoroughly affects the prior probabilities: the referring physicians will have "stopped" many non-serious cases of acute abdominal pain from entering the surgery ward. The chances of such a patient's having acute appendicitis, therefore, are much higher in a surgery ward than in an emergency department. Also, de Dombal and his co-workers encountered huge terminological differences among regions. Different groups of physicians appeared to use completely different definitions of "acute abdominal pain." Some included urology patients, others did not.[6] Similarly, in introducing the system elsewhere, new forms had to be designed, symptoms had to be carefully defined, and physicians had to be trained anew to ensure that in saying (e.g.) "the pain is colicky" they would mean the same thing that the Leeds doctors had been taught to mean. Finally, the presentation of some diseases seemed to vary between places. Pancreatitis, for example, behaved differently in Copenhagen than in Leeds, and this caused the Leeds program to falter.[7]

The non-transferability of databases, thus, is due to a complex amalgam of interrelated differences in organization, terminology, disease presentation, and so forth. Differences in opinion as to what constitutes good medical practice can also play a role; for example, Shortliffe et al. (1979)

attribute the fact that a system built by Bleich (1972) had never been in widespread use to its "'Bleich-in-the-Box' feature." In their introduction to a volume on computer-assisted medical decision making, Reggia and Tuhrim (1985) sum up this thorny issue as follows: "This is more than the usual problem of physical portability of a program: it also involves resolving differences in medical definitions, varying standards of practice, and differences in patient populations."

Non-transferable databases, however, are not the only obstacle to the implementation of a tool beyond a singular site. Other problems originate from differences in the way the tool is to be used. The following remarks, in which a resident surgeon comments on the difference between using de Dombal's abdominal pain system in a surgical ward (in Airedale) and using it in an emergency department (in Leeds, where this physician quoted here worked), point out how a myriad of detailed organizational features inhibit smooth transfer from one location to the next:

The constraints of using the system are, compared to Airedale, not as strict here. [The form] is not part of the patient file, for example. They have to separately pick up the form here; they can very well do without it. These are different worlds, the ward at Airedale, and the emergency department here. There is no one standard way of implementing the system. At Airedale, physicians have ten to twenty minutes to talk with each patient. Plenty of time. Here, five minutes, and you've got to make a decision. It's a mess here. You can't find the form, what do you do? You've got to deal with the patients; it's so much more hectic in a place like this. (July 16, 1993)

According to this physician, this tool's script fits with the way Airedale's organizational routines are structured. Since work in the Leeds emergency department is less ordered and is performed under greater time pressures, physicians are less willing to fill in the required thirty-some data items on the form.

At the time I interviewed Peter Emerson, no attempt had been made to export ACORN to another hospital, so whether this could be done remained an open question. When asked about this possibility, Emerson sighed and remarked that the situation would be very different in other hospitals: "If you would have someone who's enthusiastic for it, then maybe . . . But it would be a lot of trouble. . . . Just dropping it there wouldn't work." (interview July 21, 1993)

Wyatt (1991a) lists some reasons why exportation might be problematic. ACORN, he argues, was in large part directed at reducing delays, which might not be a problem in other hospitals. Other hospitals might be better staffed than ACORN's emergency department in London, or they might have spaces in the emergency department where more intensive supervision and treatment of possible myocardial infarction patients is already possible. Furthermore, in other hospitals more experienced medical staff members might be available to interpret ECGs, which might make ACORN's reliance on the automated ECG interpretation undesirable or unnecessary.

Similar issues affect the construction of protocols intended for use at multiple sites. A protocol requiring a certain array of tests at a certain time will not work in a setting where one or more of these tests are not routinely available, or where the organizational arrangements are such that the protocol's time schedule is unworkable. A protocol requiring a lumbar puncture (a puncture made in the lower back to withdraw spinal fluid) on "day 4" of the treatment scheme will generate problems in smaller hospitals where these punctures are not routinely done on weekends. Similarly, a demand to do "radiotherapy" at a certain fixed time might run into problems in hospitals that do not have their own radiotherapy facilities and thus need to refer patients to clinics.[8]

In the cases of interinstitutional protocols and tools like de Dombal's, then, the problem of transferability does not disappear: it is merely replaced. In these instances, the reach of decision-support techniques is limited to those hospitals or departments that are alike enough to make common use possible. The "FAC scan" needed to perform the CD34 test in the breast cancer protocol, for instance, is an expensive instrument which is (at least in the Netherlands) available only in university hospitals and some of the larger general hospitals. Even less widely distributed is the technology to perform a peripheral stem cell transplantation. Finally, the diagnostic and therapeutic concepts contained in an interinstitutional protocol might be dismissed as too academic by physicians working in rural areas, or, conversely, as too coarse-grained by physicians working in university hospitals.[9] The latter typically sneer at standards or consensus statements produced by national institutions as being for "ordinary hospitals." For them, *not* adhering to such protocols is what gives them extra status as first-rank, university-based clinicians. Localization in space, here, is not so much confinement to a single site as it is confinement to a limited series of local practices whose configurations are similar enough to allow for a single script.[10]

Constructing a feasible, workable decision-support tool, then, always implies *building specific contexts into the technique.* Inevitably, idiosyncratic, unique features of the specific sites involved become embedded in a tool's script; inescapably, in the construction and implementation of decision-support tools, a tool's radius of action is reduced.

As the example of de Dombal's tool shows, however, this type of localization can be combated—but this requires much additional work. Swedish physicians had to be trained to adapt the same terminology and investigational techniques as their Leeds colleagues, organizational setups had to be screened or altered, and emergency-department physicians had to be enticed to use the system as their surgery-ward colleagues did. Also, much larger, international databases have been gathered to ensure their validity across regions—albeit with varying success.[11] To overcome localization in space, additional disciplining of the involved medical practices to the formalism is needed. Stable relations between data items and outcomes now have to stretch out over regions. Physicians in different locales have to be trained to use the system and the involved terminology uniformly; organizational differences have to be flattened out.

Similar attempts have been made to overcome the idiosyncratic nature of local protocols: Eddy tried (with only partial success) to base his tool on international data, and several institutions have developed guidelines on the writing of guidelines.[12] It is through these processes that localization in space can be countered: only in and through this work of aligning organizations, terminologies, and procedures does a tool acquire its potential to work everywhere.

On the other hand, everywhere is always somewhere: the built-in context always locates a tool in (a) specific place(s). Medical procedures of research protocols embed the bureaucratic structure of modern hospitals: chemotherapy courses are always repeated at intervals of seven, not nine, days; infusions are given over 24 hours, never over 27. Also, when the required disciplining fails, localization in space is one way to cope with the failure. There were many instances where de Dombal's team did *not* succeed in adjusting the practices to the requirements of the widespread use of their tool. In messy and complex practices such as hospital wards, control is never complete. Nurses, physicians, patients, and hospital administrators all have their own tasks and agendas, to which the decision tool might be accommodated. The concerns of medical personnel for what *they* think should be done for the patient might overrule their allegiance to the tool, or they might see the tool as imposing too many additional duties on them. Staff members' conformity to the tool's demands might be too hard to obtain, just as the controllability of data items and the patients' afflictions is sometimes beyond reach. In those cases where this unruliness could not be overcome, de Dombal and his co-workers went back to their system and tinkered with its setup so that these problems would not result in the total breakdown of the project. Whenever they did not manage to flatten out the organizational differences between different locations, for example, they just created a different tool for each site. Likewise, when builders of the breast cancer protocol were not able to prevent some institutions from giving their own local protocols higher priority, they just reduced the number of sites where the protocol would be used.[13]

Tools can also become *localized in scope*. This occurs when the field of application of a system is limited to an increasingly small subset of tasks. This may mean reducing the range of disorders the system can handle, or limiting the scope of actions it can undertake. In ACORN's practice, the designers did not succeed in incorporating non-cardiac causes of chest pain. Also, the tool had to be made "less intelligent": the thirteen detailed advice options were cut down to three. Here again, ACORN is quite typical of the overall category of computer-based decision tools. Within limited fields, tool builders often say, their tools perform well. "Skilled behavior" is seen in "many relatively well-constrained domains." However, they then continue, most programs exhibit the "plateau and cliff effect": "the program is outstanding on the core set of anticipated applications, but degrades rather ungracefully for problems just outside its domain of coverage" (Szolovits 1982).[14] Increasing the scope leads to problems: unforeseen interactions between rules appear, or the system loses precision or fails to function altogether. De Dombal's team repeatedly tried to enlarge the scope of their system beyond the small domain of acute abdominal pain: they attempted to include, for instance, gynecological disorders, dyspepsia, or Crohn's disease. None of these attempts were very successful: the system would always perform much better with the added categories left out.[15]

Like localization in space, localization in scope is a result of failing control. It is a matter of unruly, heterogeneous elements that cannot be brought in line with the system. It is never "just" a "technical" matter of the "broad and complex" nature of clinical problems, as builders of decision-support tools often state (Szolovits et al. 1988; cf. Forsythe 1993b). It is not that ACORN failed in the realm of non-cardiac causes of chest pain because of the "inherent difficulty" of that problem. The situation is both more complex and more mundane. If ACORN's designers had succeeded in arranging for nurses to elicit additional data items, such as an x-ray, blood tests, or auscultatory sounds, it might have been possible to create a set of data items that could be used to distinguish among several pulmonary afflictions. This was not feasible, however: the physicians would not let nurses do these tasks, and the nurses did not want to do them much either. Also, it would have taken additional time to gather these items, and it would have generated new problems—such as how to turn x-ray interpretations (which are notoriously vague) into clear-cut, unequivocal pieces of input.[16] All in all, the failure to increase the scope of a tool is thus as much a failure of social control as it is a problem of the intractability of the relations between data items and disease categories. Including pulmonary diseases in ACORN would have required a stronger grip on nurses, physicians, the interpretation of x-rays, and the links between test data and their implications than was feasible in this place at this time. In this sense, ACORN's localization in scope was not so much due to a mysterious "inherent difficulty" in the nature of chest pain as to a practical failure to adequately constrain the closely intertwined, heterogeneous elements of this emergency department.[17]

Similarly, the authors of the breast cancer protocol first included all patients with surgically removed breast cancer—whether they had undergone a "radical mastectomy" (removal of the whole breast) or a "breast-conserving procedure" (where the tumor is excised but the breast is not removed). The radiologists of one of the participating centers, however, did not agree with this. These physicians felt that patients who had had the breast-conserving procedure required immediate radiotherapy to kill any remaining tumor cells in the breast. In the proposed protocol, radiotherapy would come at the end, being delayed some 16–18 weeks (see figure 3.1). Although not all radiotherapists felt that this radiation of the breast was so urgent, the authors of the protocol felt obliged to give so as to retain the cooperation of these radiologists. Consequently, the next version of the protocol excluded patients treated with a breast-conserving procedure.[18]

The third and last type of localization I want to discuss is *localization in rationale*. This occurs when designers must agree to a setup in which the potential, ideal-typed precision, accuracy, or "scientific quality" of the system is not realized. As in the previous types of localization, localization in rationale inevitably occurs as a result of the ongoing negotiations during the development and implementation of a tool. Faltering attempts to adequately discipline parts of the practice may be managed in this way too. This comes out clearly in the account of Eddy's guideline: in order to keep the guideline feasible, Eddy had to swallow so many compromises that the guideline "seemed crazy." As Thomas Bear (an oncologist involved in the writing of the breast-cancer protocol) put it, protocols are always "a political compromise." They will always have to be tinkered with to be not too different from what physicians are already doing, and from what is organizationally practicable. Wiersma (1992) describes how the "working group" responsible for the Dutch general physicians' consensus report on serum cholesterol tests and treatment continually struggled with these issues. The working group, for instance, found that high levels of serum cholesterol are much less strongly correlated with coronary heart disease for women than for men. But, Wiersma says, having "different guidelines for screening women was felt to be both socially unacceptable and impossible for the practitioners to advocate to their patients" (ibid.). Finally, Komaroff et al. (1973) argue that developing a protocol also means deciding which patients the protocol "might be expected to miss." In the case of the protocol for upper-respiratory-tract infection, they state, it would have been possible, in principle, to check every patient for asymptomatic otitis media (inflammation of the middle ear). Doing so, however, would have required looking in all patients' ears—an activity usually considered to be too difficult for physician extenders to perform. Wanting to find these patients would thus have required dropping the whole idea of the physician extender, since the physicians would have to see each patient anyway. As a compromise, Greenfield's protocol simply asks the patient about ear ache. By missing the asymptomatic otitis media cases, by reducing the protocol's potential precision and accuracy, they saved their project.

Similar phenomena occurred in the case of ACORN. When Emerson started out building his tool, he was "a devoted Bayesian" interested in rationalizing medicine—in the "mathematical approach to decision making" (interview, ibid.). Already when selecting the items, however, Emerson and his co-workers had to give up some of this ideal. Potentially powerful data items had to be left out because the nurses could not or would not elicit them. This "corruption" of the Bayesian ideal grew worse when they had to incorporate a wide variety of symbolic, *ad hoc* rules to prevent impermissible actions (such as sending home patients who looked gravely ill). In the end, Emerson and his co-workers abandoned the statistical ideal and embraced a clinical logic. ACORN in its final version is a rule-based expert system in the classical sense. Judged from this clinical ideal, however, the tool's rationale was also limited: too many data items with potential clinical value had to be left out to keep the tool workable.

The ideal of statistical inference was also compromised from the outset in the case of de Dombal's tool. Its limited scope of six or seven diseases led to the requirement that the physician make the first selection: Is this patient suitable for the tool at all? Does this patient belong to the category of "patients possibly suffering from acute appendicitis" (Horrocks et al. 1972)? This necessity for the subjective judgment of the physician is, of course, a pollution of the purity of the ideal-typed statistical logic. Also, over the years, de Dombal's tool has evolved toward a rather different philosophy. In recent articles, it is continually stressed that the tool is *not* a decision maker. It is "just a test," or even only an educational tool; it yields nothing more than an indication as to what might be the case for a given patient (de Dombal 1991; de Dombal et al. 1991). The decision itself is returned completely to the hands of the physician. From the perspective of the original Bayesian ideal, this is a complete capitulation: an invitation to let the arbitrariness back in.

A similar movement can be said to characterize the whole domain of rule-based expert systems. Many authors are abandoning the traditional Artificial Intelligence approach in favor of building tools that function more like electronic textbooks. The time of the "Greek Oracle model," Miller et al. (1986) argue, is over. No longer should we work on computers employing "seemingly superhuman reasoning capabilities to solve the physician's diagnostic problem" (Miller and Masarie 1990), with the physician merely a "passive observer" (Lipscombe 1991).[19] Here, too, a step back has been taken from the claim that the tool outperforms the physician at the moment of decision. And here, too, the original ideal of mimicking and even ameliorating expert reasoning has been thoroughly watered down.[20]

The localization of a tool is inextricably tied up with the disciplining of a practice to that formalism. In the process of getting a decision-support tool to work, both phenomena inevitably occur. The medical practice rewrites the tool's script and vice versa; the final result is a configuration in which each has transformed the other to its own liking. The various types of localization (in space, scope, and rationale) can be fought through trying to strengthen control, through tightening the network of nurses, patients' bodies, physicians, ECG machines, and so forth. As has been said before, the limits on the "formalizability" of a medical domain are not a pre-set given: they constitute a moving frontier, a place of struggle for control. Moving the frontier involves an exponentially increasing amount of work: Eddy and his colleagues could have fought the corruption seeping into their guideline by doing the time-and-motion studies to estimate costs, or by performing trials to determine the risks of adverse reactions. On the other hand, difficulties in adequately constraining the heterogeneous constituents of medical practices can be coped with by localizing the tool. In a practice disciplined to a formalism, what is "legal" and what is "illegal" behavior is clearly distinguished. Only some movements are allowed by the script; as Bowers (1992) phrases it, only "some orders or compositions can be realized or recognized in a particular formalism." When resistance occurs, or when "illegal" conduct (whether of people or of laboratory tests) is too frequent and is not amenable to correction, the tool can be tinkered with to save it from failure. By incorporating "craziness" into the tool (the final guideline agreed only in part with the decision-analytic calculations), Eddy kept himself from burning out. By reducing the scope or the rationale of the tool, or by accommodating local idiosyncrasies in the tool's design, a project can survive.

The various types of localization are not meant as an exhaustive set of mutually exclusive categories. Nevertheless, they allow us to view the development of both individual decision-support techniques and classes of such techniques as attempts to avoid a type of localization. These attempts often allow another type of localization to come in through the back door. More often than not, for example, the widespread distribution of a protocol is "bought" by reducing its specificity. As Bowker and Star (1994) have argued in the case of disease classifications, "standardization-procedures must be tailored to the degree of granularity that can be realistically achieved." A certain, necessary "uncertainty and ambiguity" (ibid.) is needed—and more so where more locales or specialties have to be included within one "universal" protocol. And, indeed, authors involved in the making of nationwide consensus statements or "standards" speak about the need to build "sufficient space for individual variation" into such tools (Casparie and Everdingen 1989; see also Kanouse et al. 1989 and chapter 2 above). Only by reducing the script's tightness can such protocols ever hope to achieve widespread use. Similarly, the oncologists working on the breast cancer protocol decided in an early stage to give recommendations for the radiotherapeutic treatment but to include in the protocol a statement that modifications of the radiotherapy according to local views are permitted. They succumbed to their conviction that they would never be able to get the radiologists in the various centers to agree on one set of prescriptions.

These examples illustrate how inscribing vagueness into the protocol can lead to more widespread use, and how localization in space can be fought by increasing localization in rationale. In contrast, ACORN's designers, in statistically pruning the list of items generated by the physicians, can be said to have traded localization of its rationale for localization in space. In order to increase the predictive power of the data items used, the designers utilized a statistical technique that took into account only a very specific population: patients with chest pain coming to a specific emergency department. Thus, while they increased the accuracy of their tool, they simultaneously tied it more tightly to the local context of the specific emergency department they were dealing with.

Changing Practices: Decreasing Diversity?

In chapter 3, I argued that disciplining a medical practice to a formalism tends to transform medical criteria in distinctive ways. The predisposition to build simple, robust worlds results in their association with a limited set of simple, clear-cut variables. Medical criteria also change in yet another way. To bring this into view, let us take another look at the oncologists' discussion of the breast cancer protocol presented in chapter 3.

At the end of their deliberations, the oncologists settled on four positive nodes. To reduce the large number of eligible patients this criterion would yield, they reduced the upper age limit from 60 to 55. Also, the originating center had set a maximum time of 14 days between the surgical removal of the tumor and the start of chemotherapy—a tradeoff between the time required for the patient's recovery from surgery and the need to prevent potentially remaining cancer cells from proliferating. Other centers, however, objected to this, arguing that this time span would be too short to allow them to obtain patients from regional hospitals. One oncologist argued: "If we want their patients as well, and I think we do need them in order to reach our quota, then fourteen days is too short. I mean, you'd have to wait for the regional oncology meetings; that's how it goes. And we have these meetings twice a month." Finally, they settled for six weeks.

How do these observations relate back to the hopes and goals of the tools' advocates? The contingent nature of the protracted process of negotiations, I argue, is incorporated in the core of the tool. Trying to get a protocol to work is a process of making *ad hoc* compromises, going back to the tool, and tinkering to get the medical practice's elements in line. The number of lymph nodes is juggled so as to articulate the heterogeneous issues involved: the tool makers take whatever opportunities they perceive in order to adequately constrain the links between the diverse, constituent elements of the medical practices. The protocol's "final" state is the highly *contingent* outcome of all this tinkering. Radiologists were given their way, entry criteria were modified, and so forth. Continually, contingencies erode the adjustments accomplished; continually, idiosyncrasies seep in from all sides. This unending, *ad hoc* compromising leads to a tool nobody had planned beforehand.

And because we are talking about decision-support techniques, we are talking about embedded decision criteria nobody had planned beforehand. We are talking about the number of lymph nodes or the age limit used to determine entry into a treatment scheme, or about the amount of drugs administered, or about the type of radiological contrast agent used (as in Eddy's account). We are dealing, thus, with transformations in the *criteria* used to enroll some patients in an intense treatment schedule and to exclude others, and with changes in the treatment given—with, therefore, what "untreatable" breast cancer *means*, and what proper treatment *is*. There is no straight line between the first blueprints of the tool and the tool the practice ends up with. There is, similarly, no simple "ironing out" of idiosyncratic variations. Instead of changing a substandard practice Y to an optimal practice X, we end up with practice Z: a hybrid containing traces of X, Y, and all the struggles that have been fought. Instead of the transparent, optimal, unified Clinical Rationality hoped for, we end up with opaque, impure, *additional* rationalities. Instead of imposing order where there was disorder, an order is achieved that *incorporates* the very messiness it started out to curtail.[21]

Similar conclusions can be drawn regarding the computer-based tools. In the case of the early Bayesian version of ACORN, we saw how the creation of the data item list began as a list generated by physicians, which was subsequently cut down by checking the individual items' reliability in the hands of nurses and by statistically determining their discrimination power. Through negotiating processes between nurses, physicians, the physical capabilities of ACORN, the possibilities of the statistical techniques, and the patients' hearts, a small list of items, with their mutual relations, was obtained. The knowledge incorporated in the practice including this new tool, thus, was a compromise among the needs of the heterogeneous elements involved. Seen from a "purely" statistical point of view, the end result would have been a practice comprising highly corrupted decision strategies. The ongoing negotiations during construction and implementation stripped all purity away: the statistical techniques were not invoked until the list of data items, created by questioning the clinicians, appeared too large. Moreover, the list of data items was cut down further by a very pragmatic requirement: the repeatability of the data item in the hands of nurses.

From a "purely" clinical point of view, the situation is equally incomprehensible. The construction of ACORN's later, expert-system version started out with a reduced version of the data item list described above, and worked from there. Thus, it began with building blocks that already formed an impure mixture of statistical and clinical logics. Moreover, the endless *ad hoc* cycles of refining the rules, debugging unwanted interrelations between them, and altering the advice options distanced the system further from a simple "simulation" of physician's reasoning.[22] As the end result of the complex negotiations in the implementation of computer-based decision tools, we encounter practices that embed decision strategies not known to or fathomable by any physician. Questions are omitted which are deemed essential by physicians (e.g., questions about the nature of the pain), and other data items take on relevancies unintelligible to doctors.

Thus, the work of getting a computer-based decision tool to function alters knowledges and decision strategies beyond recognition. The ensuing logic of ACORN's practice is of a new, irreducible kind: clinical and statistical logics are mingled with each other and are infused with a host of *ad hoc* alterations resulting from resistances encountered in the medical practices involved. This generation of idiosyncratic, impure logics, aside from being an intriguing phenomenon in itself, runs counter to the hopes and goals of those who want to cleanse medical practice of its variations. Rather than cleanse medical practice of diversity, ACORN introduces an approach to patients with chest pain equivalent to that of no physician or practice anywhere else. Stated somewhat more generally, computer-based decision techniques do not do away with the heterogeneity of logics in medicine. Variations in medical practice will not be abated through these tools. As in the case of protocols, they do not bring the sound, unitary Logic of Science hoped for. They incorporate the very idiosyncrasy the tools' advocates wanted to expel from medical practice. By introducing new logics and new rationalities, they continue, or maybe even intensify, the multiplicity they set out to erase.

Some Conclusions

The creation of a niche in a local practice for a decision tool, I argued, can be understood as a dual process of the disciplining of a set of heterogeneous elements constituting the local medical practice and, simultaneously, the localization of the tool. Three recurring ways were distinguished through which the required disciplining is often attempted: reinforcing bureaucratic hierarchies, materializing the tool's demands, and shifting decision power to the most uniform and predictable elements. Neither of these "strategies" is foolproof, however: hierarchies and forms can be circumvented or resisted, and seemingly stable elements can be impossible to gather. Building simple, robust worlds is a hard task indeed. The tool and the practice transform each other to each other's image: the practice is redesigned to mirror the tool's formal structure, and the specific contexts from which the tool emerges are inscribed in its design. Drawing upon a notion coined by Bowker (1994b), I call this the *convergence* of tool and practice.

The processes of convergence lead us into a world that was not there before. The tool does not simply come to mirror some optimal or preexisting practice, nor does the practice come to wholly reflect the tool's original script. On the contrary, the tool's formal setup restructures the knowledges it carries in distinctive ways. Similarly, the knowledges become "corrupted" with contingent cutoff lines and idiosyncratic considerations through the *ad hoc* tinkering that is part and parcel of getting a tool to work. Through the mutual redescriptions, the medical knowledge embedded in the tool is inextricably interwoven with its "context" (cf. Akrich 1992). The decision technique inhabits this new world as an active entity, but the tasks it performs are not simply "tasks taken over" from the medical staff. In this taking over, the tasks are fundamentally transformed. This issue, again, generally eludes both advocates and critics, who remain stuck in arguing which human tasks a formal tool can take over and which it cannot. In thus reifying the debate, they render the question meaningless; in equating the tasks, they ignore the crucial transformations that are taking place before their eyes.

Two final points remain. First, I want to point at a particular paradox confronting the statistical tools and expert systems. These tools, I argued in chapter 2, contain an ideal of rationalizing medical practice through single, discrete interventions in the work of medical personnel. At the moment of decision, some type of mechanical inference (statistical or symbolic) should take over from the physician and replace or support the physician's imperfect cognitive capacities. Whatever happens "outside" this cognitive realm, such as the physical gathering of data or administering of a selected therapy, is of secondary importance and need not be tampered with. In fact, it is an important claim that these issues *should not* be tampered with. Builders of decision-focused tools criticize the protocol for striving for standardization for its own sake. Rather than constrict the physician in bureaucratic webs, these designers want to intervene only and superbly at the single moment when they are most needed: the moment of decision. The builders of expert systems phrased this desire explicitly: ideally, these authors hoped, the rules within a tool would become so fine-grained and so encompassing that the tool would smoothly slide into the practice, its input and output features articulating effortlessly and powerfully with the ongoing activity.

This, however, has remained an aspiration and has lost most of its luster. The desire to create formal tools that hide their formality is like the desire to create light without heat.[23] The introduction of decision-focused tools inescapably entails keeping the heterogeneous elements of the involved medical practice in check—including increasing restraints on the actions of nurses, physicians, and patients. Moreover, this disciplining is not confined to the direct dealings with the tool: laboratories that supply data can be involved, the whole encounter with the patient can be transformed, and so forth.

The fact that a decision-focused tool's logic is more often than not fundamentally unintelligible to the physician dealing with the tool only strengthens this phenomenon. Confronted with such a tool, the physician really has no choice but to either accept the tight lead offered by the tool or to abandon the tool altogether. An expert system or a statistical tool often simply does not allow for short-cuts and unforeseen situations. The physician is presented with a pre-fixed series of questions to which he or she has to enter pre-defined inputs. And if, for example, the physician can get away with not answering or partly answering a question, he or she has no way of estimating the consequences. The tool's rationale is hidden within a literally opaque black box. Moreover, the whole process of interaction with the tool is unidirectionally geared toward the "single-moment intervention," and the "plateau and cliff effect" means that any deviation in the route toward this moment might inadvertently push the system off course.

Paradoxically, therefore, we are witnessing tools that, in order to generate a one-moment intervention, require the disciplining of a much broader realm of activities. One could say that their failure to live up to their ideal projects outward into the surrounding practices. To compensate for the inner limitations of these tools, a tight hold on their environment becomes necessary to maintain their functioning. And as the localization of the tool's rationale renders its interior more and more obscure, this hold inevitably becomes firmer. The decision-focused tool becomes somewhat like the Trojan Horse: while pledging to turn medical practice into a science, while hoping to free medical practices from the "encroaching control" from outside the medical profession, the tight network it necessitates will coalesce with this control rather than detain it.

A final corollary of the scope and depth of the mutual transformations I want to point at is that decision-support tools are, and are perceived to be by the personnel affected, thoroughly *political* tools. The scripts of decision tools inevitably embed prescriptions as to who is in charge, and who gets to use which technologies (and thus gets the status that goes with that).[24] In prescribing expensive drugs and tests, the tools pre-define where what costs are incurred, who gets to make these decisions, and who is accountable to whom. Similarly, the tool redefines which laboratory tests are important and which medications are used—and, therefore, whose labs are passed, and whose offices are filled. Physicians working with research protocols are often "wielders of hope": in their hands lie new drugs that might help where other medications have failed. Furthermore, the tool determines which patients are eligible, how much risk and suffering are tolerable, who is "high-risk" and who can wait in queue for the doctor, and who gets "a last chance."

Contrary to the rhetoric of their advocates, the tools do not confine themselves to carrying rational knowledge. Rather, their scripts redistribute responsibilities and redirect patients. Political choices, therefore, are inescapably intertwined with the "knowledges" carried. Bringing such choices into sight is especially important since the way these tools are presented hides these implications from view. The emphasis on their rationalizing potential—the "scientific authority" with which their discourses endow them—diverts attention from these very real consequences. The seemingly straightforward legitimation of being part and parcel of Rational Medicine can destroy the memory of what was needed to make the tools' functioning possible (Bowker 1994a).

These politics are just as much an unplanned, contingent result of the negotiation processes and an unexpected consequence of the formal characteristics of the tool as the decision criteria embedded in the tool. The precise responsibilities of nurses as set by the final version of ACORN and the leeway radiotherapists would get in the breast cancer protocol could not have been predicted beforehand. The politics in the lines of the protocol, or within the wires of the computer-based decision tool, cannot be directly mapped onto the intentions or interests of the tool builders.

Focusing on the politics of individual tools is much more useful than debating global questions such as whether formal decision tools are "deskilling" or "empowering" and whether they are "dehumanizing" or "restorers of hope." The sheer complexity and scope of the negotiation processes changes local networks thoroughly—often in unforeseen ways. The way to "judge" individual decision-support tools is to look at how they change individual elements' juxtapositions with other elements: Do the reshufflings leading to the silencing of some voices outweigh the increases in responsibility elsewhere? Is the way patients are described in the script of one tool preferable to the way they are described in the script of another? To whom happens what? What capacities are attributed to what element? Getting to the politics implies scrutinizing the way the new local network's relations redefine its constituents: Do the nurses' newly gained tasks outweigh both the tight lead ACORN imposes on them and their transformed relations with the emergency ward physicians? Do the protocol's criteria include too many women in order to ensure a sufficiently large flow of patients? When the tools' straightforward universality is a chimera, we cannot expect our global judgments of them to carry very far either.

5
Supporting Decision-Support Techniques: Medical Work and Formal Tools

In the previous two chapters I looked at how tool builders reshape practices and tools in the process of getting their decision techniques to work. In this chapter, I shift to studying medical personnel at work. The focus turns and centers on the primary work tasks of those who were so far just an element in the heterogeneous networks that had to be disciplined.[1] How do formal decision tools figure in the work of medical personnel? How do they transform this work? Should we conclude from the previous chapters that a functioning decision-support tool implies that the medical staff is rendered fully docile? And if not, what does the resulting configuration look like?

To get a grasp of these issues, the chapter begins with a rather basic question: What is medical work?[2] I question the views of medical practice contained in the decision tools' discourses. I challenge, more specifically, those aspects that rhetorically allow for a smooth fit between medical work and the functioning of the formal tools: the clear-cut nature of medical data and criteria, the separation of the "scientific" content of medical work from its potentially biasing social context, the stepwise nature of medical practice, and (for the decision-focused tools) the prominent role of the physician's mind. By attributing these characteristics to medical work, the discourses discussed in the first two chapters create a seemingly perfect match between medical practice and the formal and "scientific" nature of the tools. In these ideal-typed views (modeled after positivist images of science and the developing tools themselves), the nature of medical practice inherently provides for the requirements of the formal tools.[3]

This now-severed "natural link" is recreated in concrete practices through the disciplining of practices to specific formalisms. The entry of these formalisms as active agents into these work practices has important consequences. The newly configured practices transform medical personnel's work tasks, may open up new dimensions in their work, and articulate new relations—both within and outside the concrete medical practice involved.

At the same time, however, studying medical work shows how formal tools can function without a full disciplining of the practices involved. In this chapter I will argue the seemingly paradoxical claim that a practical way to get formal tools to function is by *not* transforming medical personnel into fully docile bodies. In and through their work, medical personnel can actively produce many of the tool's prerequisites while partially reappropriating the tool in the process. Only in this way, it will be shown, is the transformative potential of the tools realized.

These arguments have consequences for some of the main claims of both critics and advocates of the tools, which will be discussed in the final sections of this chapter. Through "pre-analyzing" the decisions, and through taking some responsibility away from the physician (whether only at the moment of decision or during a whole sequence of actions), the work of medical personnel was supposed to be made easier. But is this the case? How does the need to actively produce the formalism's requirements affect the advocates' hope of simplifying a practice through decision-support techniques?

On the other hand, many critics of decision-support techniques have argued that these tools take control of medical decision making out of the hands of the physicians. In their view, this control should be returned to doctors, either through abandoning the tools or through fundamentally changing their design.[4] These critics, however, overlook that these formalisms actively transform personnel's work through the medical personnel's production of the formalism's requirements. Neither the tools nor the personnel can be said to be in control. Rather, control is distributed among them in intriguing and poorly understood ways—and it is in this distribution that the power of the resulting hybrid lies.

Medical Work: The Heterogeneous Management of Patients' Trajectories [5]

I met Mr. Wood during my participant research in the oncology ward and the outpatient clinic of a university hospital in the Netherlands. Two years earlier, his internist had told him that he suffered from Hodgkin's disease: cancer arising from the lymphoid tissues. He had visited his doctor because of his persistent coughing, his fatigue, and the fact that his weight had dropped, in only a few months, from 120 to 100 kilograms. The internist found tumorous tissue in the neck, in both armpits, and in the mediastinum (the space between the lungs). Mr. Wood was given chemotherapy, followed by radiation of some of the larger tumors. His disease responded well: after this treatment, his symptoms were alleviated and the tumors disappeared.

Some months later, however, Mr. Wood was back at his internist's office. He was gradually feeling tired again, and his weight was dropping. What he had been afraid of appeared to be the case: his Hodgkin's disease was back. The cancer, as all too often is the case, had not been eradicated completely. He was sent to the university hospital to see whether he would be eligible for bone-marrow transplantation. In this therapy, bone marrow collected from the patient's pelvic bone is deep-frozen. Subsequently, the patient is treated with massive doses of chemotherapy. These doses are so high that they kill the bone-marrow cells, responsible for the production of blood cells. Without these cells, vital parts of the immune system are depleted and the patient dies of infection. Also, the lack of blood platelets, crucial for hemostasis, may cause lethal internal and external bleedings. In bone-marrow transplantation, however, the unfrozen bone-marrow cells are given back to the patient after the highly toxic chemotherapy, which is hoped to affect the tumor cells in equally deadly fashion.[6]

Mr. Wood was found eligible for bone-marrow transplantation and was hospitalized. When I met him, he had already had his treatment, and his bone marrow had been reinfused. This stage is a crucial one, since the effects of the chemotherapy on the bone marrow are starting to show and the "reinstalled" bone marrow is only starting to proliferate. Blood cells are counted daily, and their decline is closely monitored. The patient is investigated daily for signs of infection or bleeding. To minimize the risk of infection, the patient is kept in isolation in a special room with an air lock and double doors. To enter, visitors and medical personnel have to wear special robes and masks. The isolation procedures are continued until the blood-cell counts have again risen above a certain threshold. Then the patient is declared "out of the dip" (referring to the dip in the curve of the blood-cell-count graph in the patient's record), and the isolation measures are stopped.

It is Monday. Dr. Howard, head of the department, has been on duty this weekend. When he meets John, the senior resident who runs the day-to-day affairs of the oncology ward, he tells him to keep an eye on Mr. Wood's temperature. "They called me to say that it rose above 38.5 this weekend," says Dr. Howard. "I said that they should check again two hours later, and if it was above 38.5 again we would have to start with antibiotic treatment. But it wasn't. You've got to check it, John, because he's got 0 granulocytes [a type of white blood cells] at the moment, and that means he's prone to blood poisoning, and then they're dead before you know it."

During the day, Mr. Wood's temperature remains below 38.5°C. John investigates but cannot find a reason for the fever. That night, however, the patient's temperature rises again: when it is measured at 10 P.M. it is 38.8°C. According to her routine, since the temperature is above 38.5°C, the nurse calls the resident who is

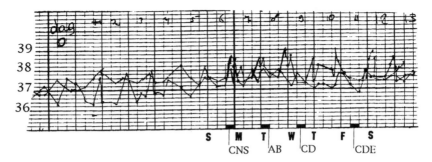

Figure 5.1
Mr. Wood's temperature curve. The relevant line is the one with the high peaks; the other represents the pulse. The vertical lines indicate eight-hour periods. S, M, T, W, T, F, S: days of the week. Numbers at top: days after bone-marrow reinfusion. AB: start first antibiotic. CD: shift to second antibiotic. CDE: third antibiotic added. CNS: time of taking of blood culture that was positive.

on night duty and sends a sample of Mr. Wood's blood to the laboratory for a blood culture. The resident, in turn, calls Dr. Howard, who tells him to wait two hours and measure again. At midnight the patient's temperature is 38.6°C. Dr. Howard is called again, and he states that this could imply that Mr. Wood has blood poisoning. Dr. Howard orders a broad antibiotic therapy, consisting of two antibiotics which I will refer to as AB. When John arrives in the morning he reinvestigates Mr. Wood and finds a small anal fissure close to a pile. He jots this finding down in Mr. Wood's file and adds an exclamation mark: this could be the source of the infectious trouble! He orders an x-ray of the lungs to search for other possible focuses, but the x-ray, according to the report, is "clean."

The next day, Wednesday, John discusses Mr. Wood's case with Liston, the oncologist who supervises John's work on the ward. "He's been having a fever for some days," says John, "and lately he stabilized above 38.5°, so we started with AB. And we've just received the blood culture from Sunday, which says that in one out of the four tubes of blood a coagulase-negative staphylococcus has been found for which AB is effective. So we're on the right track." "Yeah, well," replies Liston while she is pointing to the temperature curve, "but the temperature is still rising, so we're going to shift to CD [a different combination of antibiotics]." John is silent, while Liston writes this change in Mr. Wood's record. "CD hits that bacteria even harder," she adds.

Two days later, Friday, during the "paper rounds," John recounts this story to Bear, another oncologist who has taken over supervision duty. That night the temperature had risen again, to 39.0°. The resident on duty had not undertaken any specific action, and had not called Bear—it had been a single peak. "He is some thirteen days after transplantation" says John. "He has acquired a fever, since, ac-

tually, this weekend. We started with AB, and changed to CD on Wednesday." Matthew, a nurse attending these rounds, adds: *"Yes, and his blood pressure is low, with a pulse of. . . ."* Bear interrupts him, and continues on the topic of the antibiotics. He has not heard about this patient for a while, and he wonders about the quick change in therapy. He asks the bacteriologist, who is frequently present at these rounds, whether he has seen a positive blood culture. *"No."* Bear then sees in the record that one out of four tubes was positive. The bacteriologist shakes his head: *"One out of four means nothing. I would define that as a contamination, as a skin bacteria which has accidentally entered the needle when the blood was drawn. I would see that as a negative blood culture."* John intervenes to defend the shift in antibiotics, which now looks somewhat dubious: *"Since the temperature was high and not responding to AB, we changed to CD."* The bacteriologist shakes his head again, and repeats *"One out of four tubes is nothing,"* implying that there was no good reason to shift from AB to CD. Bear now comes to the aid of John and Liston's: *"Yeah, well, we know that, but we had to shift the antibiotic treatment anyway, since the temperature wasn't reacting properly."* John, shifting the discussion somewhat, adds: *"His pile presents a similar problem. Do you have to gear your antibiotics toward that, or toward the positive tube?"* The bacteriologist is confident: *"In a situation like this the clinical situation should prevail. That's what I think."* He explains that DE would have been a better choice: that combination would have been targeted more directly against bacteria that may be present in the pile. Bear joins in: *"Yeah, that pile problem. Why haven't I been called last night? There was a high temperature, and we should have acted in some way. And what do we do now? Do we change from CD to DE or add something to CD?"* The latter is not the most elegant solution: these three agents overlap significantly. *"However,"* the bacteriologist remarks, *"you have, in fact, already made the decision. I think it would be best, now, to add E."* Which is what happens.

Some days later, Mr. Wood's temperature quiets down and the isolation measures are stopped. In the discharge letter, a month later, this episode is summarized as follows:

On day +8 [eight days after bone-marrow reinfusion] a fever develops. The focus appears to be a thrombosed [blood-clot-containing] pile. Empirically [on the basis of the clinical picture] AB is started. Blood cultures show a [coagulase-negative staphylococcus], reason to treat patient further with CD. Because the temperature responds insufficiently, E is added. The temperature subsequently normalized only slowly. The anal fissure eventually quieted down. No other infectious problems have occurred. . . . Patient was discharged in a good general condition [35 days after reinfusion] and went home.

■

Medical work is the continuous struggle to *make* a patient's case work: to keep a patient's trajectory "on track" (Strauss et al. 1985).[7] In this process, anamnestic information, examination results and medical criteria are not

128   *Chapter 5*

so much "uncovered" or "given" as they are continuously reconstructed.[8] In the next fragment, Daton and Beatty can be seen to construct their respective sets of medical data:

Dr. Daton, a general practitioner, is called in by her patient Mr. Porter, who, as Daton knows, suffers from chronic bronchitis and low back pain. This time Porter is complaining of "attacks of lung pain" (he rubs his hands against his chest) which come up unexpectedly and disappear promptly upon taking the "pills from the specialist."

D: *When you've got that pain, do you want to lie still or rather move about?*
P: *No, doctor, I can't lie still.*
D: *Does the pain start when you exert yourself, or when you leave the house when it's cold out?*
P: *No.*
D: *Do you tire quickly when you exert yourself?*
P: *Yes.*
D: *Does the pain start in your low back?*
P: *Yes.*
D: *When you're in bed, or also at other times, for example when you make a wrong movement?*
P: *Well, doctor, I'm mostly in bed or in my chair when the pain comes, but yes, I do sometimes make a wrong movement.*
D: *And the pain stops when you take the pills I gave you? (Daton had given him Feldene, a simple pain killer.)*
P: *Yes, doctor, but the pills of the specialist do work better, more rapidly.*

These "pills" appear to be Nitrostat, an anti-angina pectoris drug that reduces the workload of the heart by dilating the vascular system. Driving back to her office, Dr. Daton remarks that Porter is actually suffering from atypical low back pains. Nitrostat, she claims, "works" only because of a placebo effect.

According to the letter from Dr. Beatty, a cardiologist to whom Porter had been referred by a fellow general practitioner, the electrocardiogram and the cardiac enzymes were normal. According to Beatty the pain generally started in the left side of the chest. His diagnosis: "atypical angina pectoris."

Patients' symptoms take shape through the questions asked and the examinations performed. Dr. Daton pursues her questioning only on items and remarks she considers to be relevant in differentiating between low back pain and angina pectoris ("Does the pain start when you exert yourself?" "Does the pain start in your lower back?"). In this way, she elicits information corresponding to the outcome she is considering: in these questions, the typical diagnostic patterns are already embodied. Similarly, Dr. Beatty does not ask any questions concerning Porter's backaches. Not familiar with Porter's medical history, he does not consider this option.

In addition, questions like "When you've got that pain, do you want to lie still or move about?" already contain the pre-selected answers. The expected answer can be implied in the question: the reactions to the questions "the pain starts in the low back?" (Daton) and "the pain starts in the left side of the chest?" (Beatty) are both affirmative. The very dissimilar outcomes these two doctors are considering are both confirmed.[9]

Likewise, the selection and routine performance of examination procedures produces the pathological reality to be counteracted.[10] The precise history of Mr. Wood's temperature is the result of meticulously performed nursing routines: every 8 hours (and more frequently when the temperature rises), a measurement is made and logged onto the patient's chart. Also, John selects an x-ray of the lungs but not one of the abdomen, thus carving out a different space in which a medical problem could be localized. The piles and the fissure, finally, would never have appeared in the records if John had not specifically looked for them.

These data are not produced in order to create a true image of patients' bodies. For one, physicians are too aware of the rough and blunt character of the techniques and instruments available to want to strive for such an illusion. Diagnostic technologies have too many "ifs" and "buts" attached: John's x-ray will miss a small focus of infection, a "rising temperature" can be an artifact of two slightly differently performed measurements, and so forth. In medical work, moreover, such a "true picture" would not be the prime item of interest even if it could be produced. Medical work is directed at finding an answer to what Garfinkel (1967) has called the "practical problem par excellence: 'What to do next?'" In conjunction with nurses, nurses' tools, and organizational routines, physicians do not attempt to create "true" images of nature; they attempt to create a meaningful difference for the purpose at hand—a result sufficient to direct the immediate course of action (Bowker 1994b).[11]

Gathered data mutually elaborate one another (Whalen 1993): the pattern of examination data and anamnestic information lead Dr. Beatty to denote the attacks of "lung pain" as angina pectoris, which is often felt outside the heart area. At the same time, these elaborations can lead in wholly different directions: to Dr. Daton, who draws upon different background information (she knows Porter as a "back pain patient") the vague designation of the pain indicates referred back pain. Similarly, the normal ECG and cardiac enzymes (compatible with both angina pectoris and low back pain) are mobilized by both physicians as sustaining their divergent proposed solutions.

The acquired solidity of anamnestic information and examination results can continually be undone. New data can come to the fore, conflicting with the emerging pattern. Yet coherence is then often restored through *reconstructing* a piece of information: by rephrasing a question, for example, or by stating that the patient did not seem very reliable when the history was taken. John denotes the temperature peaks as "a temperature stabilizing above 38.5"; Liston, however, relabels the temperature as "rising."[12]

A prominent feature of these reconstructions is that there are no types of data that *always* prevail when they clash with others. It would be erroneous to conclude that examination data, for example, form the "rock bottom" to which anamnestic data are adjusted; or that laboratory data "win" over results of physical examinations. *All* data can be and are reconstructed. Whereas the fast response to Nitrostat would have convinced Dr. Beatty of his diagnosis, and would have countered any patient's remarks about the pain starting in the lower back, Dr. Daton weighed these contradicting pieces of information in exactly the opposite way, nullifying the effect of the Nitrostat by stating that it was just a placebo effect. There is no "foundation" to give us a final answer; no "basic ground" upon which these reconstruction processes rest.[13]

Medical criteria are reconstructed in comparable ways. Whereas the bacteriologist stated that "one positive tube is a negative," John referred to this one tube as a legitimation of the change in treatment. The patient had a fever, so *in this specific situation* this one tube was relevant. Similarly, in this ward it had become common practice to refrain from immediate intervention when a temperature spiked above 38.5°C a single time. More action than the routine blood culture would be necessary only when a repeated measurement, two hours later, would again show a temperature above this limit. In this specific situation, however, Bear disagrees. He should have been called, he argues, even though there was only one spike.

Thus, medical criteria should not be seen as fixed rules that merely have to be applied. Like medical data, criteria are readjusted to concrete situations: the rule "one tube is no tube" can continually be reconstructed. Here, also, no "foundation" can be found from which the processes of construction and reconstruction could be derived. On the contrary, when work proceeds smoothly—and when there are no conflicting data—"criteria" are seldom explicitly mentioned or referred to. In those moments, action is transparent; "what always takes place takes place." Rules or criteria are made explicit only when this smooth flow is interrupted—when it is no longer obvious what to do (Suchman 1987). As ethnomethodology has taught us, explicit "rules" are not the foundation of action: they are *post hoc*, contingent, situation-dependent *derivatives* of concrete action. They are codifications of "what is normally the case"—but, since such a codification is always partial and tied to a particular situation, it is always open for reinterpretation. Criteria are fluid *resources* for the actors involved, who can actively orient to them in order to accomplish the task at hand.[14] In the case of Mr. Wood, adding antibiotic E to C and D is "normally" not a very elegant combination. This "rule," however, ignores the possibility that the combination of C and D has been started recently. In this situation, stopping the use of C and D seems an even less elegant solution.

Medical practices, in addition, are not solely concerned with data and criteria: more "cross-cutting systems of relevance" interfere (Bosk 1979). Organizational features, the patient's needs and desires, or financial matters are seen as "secondary" or "bias" by the decision-support tool's discourses—or are simply neglected.[15] In medical work, however, such issues are inseparably interwoven with "medical" matters as decision criteria and examination data. The patient's insurance has to be in order and the ward's beds need not be overfilled: as mundane as these issues may sound, they play similar and equally essential roles. John, for example, has to gain Liston's backing for continuing the AB treatment. But John does not obtain her support, and Liston changes the treatment. Similarly, Matthew does not get the physician's support for his remark on Mr. Wood's low blood pressure. He is simply ignored—an experience familiar to nurses.

Organizational issues such as the support of colleagues, then, may transform the flow of medical work just as contradictory data can. As was the case with the data, such issues may lead to the reconstruction of other elements, as the following fragment from my 1991 participant notes illustrates:

Mr. York had recently been admitted to the oncology ward for an intensive course of chemotherapy. He had just started this treatment and, according to the treatment protocol, he was supposed to remain in his room. However, the protocol also required some x-rays and other investigations which had not been done yet. So the nurses wheeled him around anyway—and my questioning about this was seen as a flagrant example of not understanding "how these things work."

Here medical criteria were "overruled" because of the practical exigencies of the situation—the nurses had not had time to do the investigations earlier. On the other hand, medical personnel may intervene in organizational routines in order to secure a specific outcome. Bear risked a fight with the hospital's dispensary by ordering a specific medication he wanted for Mr. Wood without asking the dispensary's chief pharmacist first. Following the organizational rules, Bear said, would have wasted too much time and trouble.

To summarize: Medical work is a "molding process in which the patient and his situation are reconstructed to render them manageable within existing agency routines" (Rees 1981, quoting G. Smith, Ideologies, Beliefs and Patterns of Administration in the Organisation of Social Work Practice, Ph.D. thesis, University of Aberdeen, 1973). In and through medical work, a patient's problem (that is, whatever is perceived by a person and/or his environment as a problem for which a doctor should be consulted) is *transformed* into a *manageable* problem: one which matches existing work routines. This implies a limited set of actions perceived to be a sufficient answer at this time and place. All heterogeneous elements mentioned (data, organizational issues, medical criteria, and so forth) reciprocally shape this transformation *and* are molded in this process: John may try to obtain Liston's support for continuing with the AB combination, and Bear may try to reshape the organizational routines so that he will be called. In this transformation, aspects of the patient's history take shape or are forgotten and the patients' hopes and desires are transformed; the result is a patient having backaches necessitating Feldene or angina pectoris requiring Nitrostat. What counts as the solution of the patient's problem is a result of the *outcome* of the transformation; equally, what counts as the original problem is redefined during this process (cf. Davis 1986).

Medical Work as Distributed Work

There are so many more places in the world than a doctor's head.
—Mol 1993

It is important to focus on the specific positioning of the physician relative to the elements that feed into the transformation. It would be mistaken to conceptualize the former as the mastermind, hovering above the data produced, the prevalent criteria, and the organizational routines and putting these to work at will. The doctor is but one of the entities actively involved in "patient management": nursing and laboratory routines, forms, criteria, medical instruments, and the medical record all are involved in the management of patient's trajectories. To paraphrase Hutchins (1995): these entities do not stand between the doctor and the task, as tools or subordinates the physician uses and reckons with while plotting the overall management of this patient's infliction. The burden of this latter task is *distributed over all these entities*. They "stand with the [staff member] as resources used in the regulation of behavior" in such a way that the patient's trajectory is managed (ibid.). The organizational routines produce the temperature curves and the rows and columns of laboratory results, which are copied into the medical record so as to be overseeable in a glance. This work constitutes the very ground upon which "doctoring" becomes possible in the first place—but it also determines the world of action the physician has to deal with (Berg 1996). These organizational routines themselves perform part of "the management of patient's trajectories": the production of crucial data is delegated to the heterogeneous ensemble of nurses, blood tubes, laboratory assistants, and blood analyzers that transforms specimens of blood and thermometer readouts into numbers on a sheet of paper. Through the performance of this subtask, the physician's task becomes screening and intervening upon the produced columns and graphs—a different, simpler, and much more pre-structured task than having to consider and accomplish the gathering of primary data. Likewise, out of the hundreds of available antibiotics, the hospital's dispensary has selected a circumscribed list, with pre-set dosages and indications, leaving to the doctor the (again pre-structured and simplified) task of selecting from a few available combinations. Again, this pre-selection of medications to use, rendered durable in the dispensary's guide to antibiotics and in the simple fact that unlisted drugs are not supplied, is itself a subtask of the management of the patient's trajectory.

In a way, then, the physician is a resource in Hutchins's sense—an active element like all the others, whose combination of subtasks *only as a whole* produces the effect of "patient management." Only in combination with the written isolation procedures, with the nurses' tasks, with the daily rounds of the nutrition supervisors, with the structuring work performed by the record and the forms, and with the chores of his colleagues and supervisors does John's work add up to "rendering the patient manageable within existing agency routines." This is not to say that the physician is just one resource among others. Far from it: through the production of tables and graphs of data in the medical record, through the production of a limited set of medications to choose from, and by rendering the patient's body docile by incorporating it into a hospital ward's routines (Frankenberg 1992), the social and material organization of the hospital *produces* the physician as "central actor." The organizational routines construe a circumscribed, central space for the physician, and position the physician "on top" of the paperwork, at the position of maximum oversight. At the same time, other work is channeled away from the physician: ordinarily, questions about the patient's insurance or matters concerning the availability of beds are dealt with by nurses and by the ward's secretary. The form and the content of the physician's task, however, are reshaped through this positioning, leaving the physician with a limited and pre-constrained task. The overall management *emerges* from the interlocking of the subtasks: "an interlocking set of partial procedures can produce the overall [task] without there being a representation of that overall [task] anywhere in the system" (Hutchins 1995). The phrase "heterogeneous management of patient trajectories," then, points both at the fact that what is managed is a wide array of materially heterogeneous elements and at the fact that what does the managing is itself a spatially, materially, and temporally distributed whole.

The "manageable problems" thus constructed are always only provisional: they are "adequate-for-the-moment" (Heritage 1984). They are but temporary "crystallization points" which start to dissolve when the purpose at hand is no longer immediate.[16] Liston orders a change of the antibiotic therapy to CD, but she does not prescribe what to do if the temperature should again fail to respond. Such detailed, pre-fixed plans are useless. The achieved "fit" between the heterogeneous elements is precarious; the rough and blunt techniques and routines involved in constructing, monitoring, and maintaining the patient's trajectory are all too easily brought off course. At all times, new, contingent developments may occur, necessitating renewed articulation efforts. Organizational routines may break down, an unexpected result of a laboratory test may pop up, or a new supervisor may disagree with the policies of the previous one. In the midst of all this, medical personnel are engaged in a never-ending process of *ad hoc* rearticulations. They are continuously working, managing with odds and ends, to perform their tasks in keeping the patient's trajectory on track—while, concurrently, reconstructing its course. Nurses tinker with organizational routines to get the blood tests done on time while the patient should also be wheeled to the x-ray department. And physicians, in their turn, make do with what they encounter: working upon the data that the nurses draw their attention to, and working from the present situation to restore appearing instabilities. Both nurses and physicians, to paraphrase Latour (1988), "make plans that are constantly drifting away; [they] seize upon opportunities in the midst of confusing circumstances."[17]

In the performance of nurses' and physicians' tasks, then, a *reactive, opportunistic* stand is important. In the complex circumstances described, a crucial means of getting the work done is to react to topically salient events—to respond to "cues" that happen to be at hand.[18] Liston changes the antibiotics because the temperature curve tells her that AB is not working properly. Since John has just mentioned the coagulase-negative staphylococcus, Liston opts for C and D, the antibiotics that best affect that microorganism. The positive blood culture may be doubtful, but it is the first available cue to which Liston can orient herself. When, somewhat later, Bear worries about the anal fissure and the pile, he simply adds the antibiotic (E) brought up by the bacteriologist. A similar pragmatism is reflected in the habit of waiting 2 hours when the temperature goes above 38.5°C. This is a practical, informal routine based on the fact that isolated temperature peaks are recurrent and often innocent events. Adhering strictly to the rule would result in much meaningless effort: the physician has to be notified, blood has to be drawn and sent off to be cultured, and so forth. Waiting 2 hours is an *ad hoc* way to separate the wheat from the chaff. Anticipation of the need for continuous "maintenance work," of the always only provisional nature of courses set, is a ubiquitous feature of medical work—two hours later, the situation may be greatly different.

In this perspective there is no longer any place for a notion of medical work as consisting of single-moment, cognitive decisions. The term "decision" itself reflects a strong tendency to attribute agency to "point locations": to "the doctor," primarily, but potentially also to other involved individuals (Callon and Law 1995). Yet if we consider the management of a patient's trajectory as a distributed task, such a notion of decision starts to dissolve. Let us focus, for example, on the "decision" to add E to CD instead of selecting a less redundant combination. Tracing back the event, we do not retrieve the "decision" within the physician's head—we do not converge upon a single point in time and space where the action originated. Rather, tracing back the lines leading up to this "decision," we find that they *diverge*. The dispensary's antibiotics guide is part of the "decision," since it puts forth the few antibiotics that are available. Another line leads back to the bacteriologist, since he suggests picking E. And the doctors' working routines should not be overlooked: the fact that the temperature was still "not reacting properly" 48 hours after the previous change in antibiotics feeds into "the decision" by putting forth that *some* change in antibiotics is called for. And Liston's actions also constituted "the decision": since she had recently changed the antibiotics to CD, yet another large change was inopportune. "Decisions," then, are spatially and materially distributed: the lines traced back from the event disperse and end up in guidebooks, in different people located in different places, in the dispensary's policies, and so forth.

"Decisions" are distributed temporally, too. Lines can be traced back to last year's negotiations about which antibiotics to list in the dispensary's guide, to John's having found the pile three days ago, to Liston's actions two days ago, to the bacteriologist's remark on the spot. And "decisions" often come about incrementally. "Intentions" can become conclusions; previous steps can be undone, changed, or simply forgotten, or can suddenly become "decisions" when the setting changes. The bacteriologist's joke concerning the required change in antibiotics—"You have already made the decision"—played upon exactly this phenomenon: no such thing *had* been "decided" at all.

Yet "decisions" are constantly referred to *afterwards*. The *post hoc* attribution of "decisions" to physicians—the rewriting of the chain of distributed events into a story in which a doctor decides on the patient's signs (see the fragment of the discharge letter)—is part and parcel of the organizational production of the physician as the central actor. The ubiquitous phenomenon of *summarizing* is crucial here. At each new patient round, information from diverse sources is condensed and transformed into short statements of "what is the case." In the decursus forms, physicians condense the information gathered from tests, nurses, talks with the patient, and previous entries and create a concise statement of the "current problem" and its relevant history. When a patient's situation is reevaluated, the entries made earlier are taken as "ground" upon which the next evaluation proceeds—stripped from their situational uncertainties, the specific context in which they emerged, and so forth (Rees 1981). At the same time, these summaries are created in the light of the problem at hand: each and every time, the patient's "medical history" is selectively rewritten to underwrite and lead to the current state of affairs. Detailed descriptions sink back into pages that are no longer "actual" and are summarized in one sentence, and later in one word—and these ongoing summations construe "histories" and "futures" that are continuous with the present.[19]

In addition, these summaries are always also sources of continual and retrospective inspection and judgment of the adequacy of the staff's work (Whalen 1993). Whether written in the record or presented at a meeting, they make "what really happened" public for supervisors, for colleagues, and (when written) maybe for lawyers and government officials (Garfinkel 1967; Hunter 1991). As is the case with medical criteria, then, the occurrence of a "decision" is often constructed *post hoc* to underscore the "rationality" of events.

These features of medical work strongly predispose it to the production of "rational" and "typical" narratives in which treatments were "decided upon" and diagnoses were reached through an orderly, systematic approach (Smith 1990). In subsequent reconstructions, the fluidity that typifies ongoing medical work disappears, as do references to the roles of nursing routines, the medical record, or the dispensary in the constitution of events. The two peaks above 38.5°C first become a "temperature stabilizing above 38.5," which Bear later summarizes as "fever"—and the discharge letter says that this fever simply "developed at day +8." Retrospectively, E is "added because the temperature responds insufficiently." These continual reconstructions should not be seen as a falsification of history: they are "needed to produce an account ordered enough to enable action or to communicate what is going on" (Gooding 1992), and their specific form is due to the specific exigencies of cooperative medical work. Ultimately, they create the type of report exemplified in the fragment of the discharge letter, where almost every sentence reflects a history of repeated reconstructive work. As is apparent from the account, the causal role of the pile in the development of the fever had never been very clear. Similarly, the anal fissure had never been described as "unquiet"; it was simply the only clue the physicians had.

Making patients' problems manageable, thus, is an ongoing, active process of articulating a broad array of diverse elements (cf. Fujimura 1987). It is characterized by the smooth interweaving of "social" and "medical" issues, and by the constructed nature of medical data and criteria. It is thoroughly *temporally* structured: all activities are located within a trajectory's projected "history" and "future," both of which are continually reconstructed.[20] Medical work, moreover, is distributed work: the transformation of a patient's problem in a "doable" problem is not a cognitive reconceptualization of the patient's case, but a collective achievement of an interlocked assembly of heterogeneous entities.[21] Physicians, nurses, medical records, the dispensary guide, the pre-packaged blood tubes that ensure the proper handling of the proper amount of blood— all these entities perform subtasks that *together* result in the management of a patient's problem. No one is "in control": although the organization of medical work produces the physician as a central actor, this production is partly a *post hoc* event and partly *conditional upon* the distinct pre-construction and limitation of the physician's task.

The resulting trajectory is not a product of consciously developed plans, nor is it the result of a sequence of "decisions." There is no overall "plan": the actual itinerary of the trajectory is the emergent effect of the interlocking of entities performing subtasks. The complexity and diversity of the managed array of heterogeneous elements, moreover, necessitates continual adaptation and reaction to upcoming contingencies. The course the trajectory takes, then, is an "in-course accomplishment" (Lynch 1985), which is attributed to "decisions" only in retrospect. The trajectory is continually reset on the spot, as the outcome of the continual articulation work of physicians and nurses concerning their own subtasks and of the subtasks performed by the medical record and the laboratory routines.[22]

Formal Tools and Medical Work

How do formal tools figure in all this? How do they transform medical work? When a formal tool is introduced into a practice, a new, active agent is added to the interlocking chain of entities performing the management of the patient. This transforms both the overall task performed and the subtasks delegated to medical personnel. As the tool transforms and becomes part and parcel of a local network, a powerful hybrid emerges in which the interlocking of physicians' and nurses' work with the tool's functioning results in wholly novel workplaces. The protocol, for instance, forges articulations of activities over different sites and times. Nurses find leads on when to do which laboratory test, and when to shift from one chemotherapeutic drug to another. Likewise, through the protocol the radiotherapist can know when what is expected of her, and how her actions fit in the overall picture. Through the making of detailed "lists," actions and events in different spaces and times are brought together and coordinated (Goody 1977; Star 1989a; Callon 1991). The protocol functions as a focal point of reference—a common resource—to which various staff members refer, can orient themselves, and can find clues on what to do next.

In this way, by taking coordination tasks out of the hands of the medical staff, a protocol may increase the overall complexity of activities. The highly complicated oncological treatment schedules, for instance, are only *possible* through the protocol's core role: its *coordinating* function makes the elaborately sequenced chemotherapeutical combinations and the highly differentiated diagnostic schemes practicable. Likewise, during Mr. Wood's stay in the isolation room, many daily steps to take and tests to perform were prescribed by the local protocol on isolation procedures, which took planning and organizing chores out of the hands of the nurses. In addition, a protocol can create comparability of activities over time and place—both a core function of the research protocol and of those protocols that are intended to "erase variations" between practices. Mr. Wood's chemotherapeutic treatment (including directives for post-reinfusion care) was part of an international research protocol that performed the double work of linking this case to an international research effort *and* coordinating the combination, sequence, dosage and timing of the individual drugs. A coordinating tool, in other words, is a means to increase the complexity of a local network and/or to extend this network in time and space.

Other configurations are possible too: in the case of a tool that attempts to create a single-moment intervention (e.g., an expert system or a statistical tool), a different set of tasks is delegated to the tool. Here we witness more of a single-moment assimilation and processing of data resulting in a new piece of information or advice about the patient. The network can be extended here too, but not so much in time and space as in *depth:* what used to be handled as an array of data is now available as a single statement.[23] Computer-based statistical tools or expert systems can fulfill elaborate functions in amassing, assimilating, and continually monitoring the avalanche of medical data incessantly being produced. In this way, ACORN attempted to make it possible for nurses to decide on admittance without having to consult the physician: the tool allowed nurses to perform a task they could not perform before. Thus, like coordinating tools, *accumulating tools* can make complexity doable. Similarly, accumulating tools can be a means of rendering the processing of sets of data in different practices more alike. In the case of de Dombal's system, the tool drew on a standard set of rules to assimilate thirty-some data items into one prognostic statement.[24]

Here again it is important to look at the issue of who or what is in control. The physician cannot be said to be in control of the management of patient's trajectories, since the physician only performs the subtasks granted by the overall network. Decision-support techniques are not in control either. They are wholly dependent on the way other entities produce their input (if any) and act upon its output (if at all); they articulate some relations (as between the different timings of chemotherapeutics), but leave it to other entities to articulate the rest (as between these timings and the patient's agenda).[25] Yet they are not "subordinated" to the physician, as some critics would have it.[26] They become active entities in the management of patient's trajectories, sharing in the overall control and "decision making" capacities. They put new constraints on the subtasks of other entities, transforming them and thereby shifting the performance of the overall network onto a new plane.

This is why I introduce the terms "coordinating" and "accumulating" tools rather than speak of "guidelines" or, worse, "expert systems." The former terms are more specific about the mundane yet crucial capacities these tools exhibit in concrete practices, while not adopting the image of "centralized control" that is attached to the latter. The tools are much more powerful than critics would have it—yet in a very different way than claimed by their advocates. "Coordination" and "accumulation" refer to the subtasks that these techniques can perform without falling into the illusion that "expertise," for example, could be meaningfully searched for *within* the boundaries of a computer—or a physician's head (cf. Hutchins 1995).

The Real-Time Work of Making Tools Work

One important issue remains unresolved. The depiction of medical work given above challenges the smooth fit between "tool" and "practice" so matter-of-factly postulated in many of the writings of the advocates of decision tools. Contrary to how the decision techniques' discourses depict the practice of medicine, medical work is not characterized by clear-cut medical data and criteria, or by the physician's overall coordination and mastery, or by a fixed, stepwise sequence. What should we conclude from this? How is this line of reasoning related to the processes of disciplining and localization? If we would follow the lines set out in chapter 3, we could suppose that a working tool requires the erasing of contingencies—i.e., the complete conformity of the heterogeneous elements to the tool's script. Yet this is clearly not the image that the case of Mr. Wood presents us with—and this was a practice disciplined to multiple protocols (several of which were at work in the few days from the trajectory described). Then how can formal tools function if the demands they put on their input, for example, are not matched by the way medical data present themselves in ongoing medical work?

In these practices, I argue, we witness a typical solution to the problem of multiple and contradicting exigencies: the work of *producing* the tool's requirements in real time is often delegated to medical personnel. The tool's exigencies exert themselves on the other entities in the network, I argued above. Yet, rather than try to discipline medical practices so that these exigencies would always be met, the task of bridging the gap between these exigencies and the instances when the disciplining falls short is added to the subtasks of staff members. Wherever heterogeneous elements behave "illegally" (that is, whenever the tool's script is disregarded), medical personnel should intervene and rearticulate the tool with the ongoing course of events.[27]

To elaborate this argument, I focus on some moments in which the fit between tool and practice, construed in the tools' discourses, has broken down.[28]

Reconstructing medical data
Formal tools require clear-cut, elementary data items as input. Medical data, however, do not come pre-defined: they are actively reconstructed in the process of making a patient's problem manageable. Moreover, their "solidification" goes only as far as is required for the practical management of the problem at hand; at any new crystallization point, new articulations are readily forged. *Creating* the required, "digitized" data, then, is no easy task. Tool builders can labor to select only those elements whose behavior fits the tools' demands most closely, or to ensure standardized use of proper investigative techniques. Such disciplining, however, is often inadequate. All too often, data present themselves in ways not anticipated by the tool builders, or data items deemed trustworthy suddenly behave unexpectedly.

In a session with de Dombal's tool, for example, a physician ran into the problem that the patient had explicitly mentioned that his pain "radiated to the back." On the form used for registering the medical data (see figure 5.2), the physician could only mark some areas of the patient's abdomen. "We don't have that on the machine," he muttered, and checked the area representing the patient's flank. As another example, Hartland (1993a) describes patients interacting with GLADYS (Glasgow Dyspepsia System), a Bayesian computer-based decision tool for diagnosing broadly defined problems of indigestion. The system is designed to take a patient's medical history, analyze the symptoms, and generate a list of possible diagnoses and a management advice. Hartland describes a patient confronted with a question about the benefits of a particular diet, to which he is expected to answer yes or no. The patient complains: ". . . at the start it did good, but after that it didn't matter. But I can't get that across by saying yes or no." When questioned about whether she had had "heartburn," a patient remarked: "Well I don't know if it is heartburn. It's

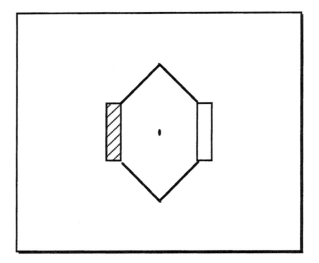

Figure 5.2
The representation of a patient's abdomen on the form used with de Dombal's tool. (For the rest of the form see figure 2.1.) The area crossed by the physician is shaded.

never been diagnosed as heartburn, I just assumed it was, I've had Brufen for it. The doctor doesn't tell me why he's given me Brufen he just gives it to me." In such situations, Hartland continues, patients often called in the help of nurses, who reassured them or helped them complete the questionnaire.[29]

Although both de Dombal's tool and GLADYS ask very basic questions, unanticipated answers immediately generate trouble. Medical data do not come readily packaged: even in recounting an elementary story of stomach troubles, unexpected translations were often required to make the story workable for a formal tool. Medical personnel have to stand by to pre-process the data whenever they escape the formats prescribed by a tool's script.

In addition, the tools discussed here embed certain translations in their rules or formulas. Inevitably, they articulate certain signs and symptoms in pre-set ways. ACORN, for example, asks whether the patient has recently experienced an unusual cough, hemoptysis (the spitting of blood), or a fever. When these and some additional questions are answered "no," the system concludes (on its printout) that "no respiratory symptom is present." This is itself an example of a translation that is "good for all practical purposes" and meaningful only in the field of operation of this particular tool. Fever and hemoptysis can both be due to a myriad of causes having nothing whatsoever to do with the respiratory tract—but the system assumes that *if* hemoptysis is present for some other reason, the erroneous conclusion that "a respiratory symptom is present" may be a bit dumb but is not consequential. In these instances, the system assumes that these oddities will be prevented or corrected by the personnel operating the tool. When a patient has had a slight fever due to an infection of his left toe, for example, the nurse using ACORN should be aware that this is *not* what "fever" means in this particular situation.

The fact that these translations are pre-set, however, guarantees a continuing tendency to stumble over contingent events not foreseen by the tool builders (Collins 1990; Dreyfus 1992).[30] The narrowly circumscribed shape and pre-defined relations of the data items in the tool contrast with the ongoing modifications and reconstructions we witness in medical work. In the case of a research protocol for patients with disseminated lung cancer, the protocol required some "routine" blood tests after each cycle of chemotherapy. These tests would indicate the general reaction of the patient's body to the therapy. Too sharp a decline in the number of white blood cells, for example, would lead to an extension of the pause between the cycles, to give the body more time to recuperate. However, with one patient who had just had her first intensive dose of chemotherapy, the nurses felt that the patient's "clinical condition" was deteriorating. She looked weaker, had severe diarrhea, and was increasingly complaining of pain. At this moment, the physician in charge ordered a CT scan, to see whether the size of her tumors was decreasing: maybe her deteriorating condition was due to the tumor's being progressive under therapy. Now, the routine blood tests mentioned in the protocol had lost their significance. *In this situation*, the implied translation "normal blood counts means everything is OK" no longer made sense. The protocol builders had not foreseen the possibility of a tumor's not even reacting to the first intensive dose—"that is a rare event," one of them later commented. The tool was not "programmed" to anticipate such early progression; the tumor's erratic behavior had disrupted the tool's links between input and output. It would have just continued on its route; the data used as indicators by the protocol were not yet deteriorating. It is the burden of medical personnel to *repair* such mismatches like this; their interference is required to untie the pre-set articulations of sets of data that have lost their relevance.

In a variation of this theme, personnel can go as far as to steer data to attempt to influence a tool's output. Doing so, they manipulate the tool's input to achieve an outcome they prefer. In the following quote a gynecology resident in a small hospital describes such a recurring situation, accentuating the pre-processing work of medical personnel with regard to medical data:

> For large infants, falling in the upper 10 percent range [i.e. weighing more than 90 percent of all newborns], we have a protocol stating that the blood sugar should be measured. We might, after all, be dealing with a missed gestational diabetes. [A large infant may be an indication of this, usually transient, type of diabetes. It is important to spot these cases because such newborns may suffer hypoglycemia, a sharp decline in their blood sugar, leading to flaccidness and even coma]. The annoying thing of this protocol is that once you're in it, you're stuck to it. I mean, Miss B.'s child weighed 4650 grams, which is just in the upper 10 percent range. But she had had a large infant before, and she's pretty big herself. She'd also been tested for gestational diabetes during her pregnancy—and all these tests were negative. So you just know that that lady doesn't have gestational diabetes, and that there is no real danger of hypoglycemia. But when you then get a blood sugar of 1.8 you're stuck to it—you can't send the baby home. The protocol tells you to keep the child hospitalized for observation: it figures that a blood sugar below 2.0 could indicate a developing hypoglycemia. They're irritating, at times, these protocols. But the thing here is that you can feed such an infant some additional glucose before the blood measurements are taken, assuring, generally, values above 2.0. And nurses know that too. . . .

Fluid criteria
Formal tools operate by processing pre-set, definite rules. In medical work, on the other hand, criteria are often fluid: they are readjusted to concrete situations, and they are explicated only when the flow of work is halted. Such explications, moreover, are always situation-specific: their relevancy for a current situation can and often has to be reassessed. In disciplining a practice to a formalism, through for example implementing bureaucratic hierarchies or materializing the tool's demands in artifacts, tool builders may attempt to fix these assessments and broaden the range of unequivocal relevancy of a tool's rules. However, these attempts too either fall short or are frustrated by fortuitous events. The codifications inscribed in decision-support tools, then, may cause at least two types of problems.

First, the tool may apply a criterion in a situation where the rule would typically be subordinated to other exigencies at hand. In the case of a protocol requiring a complex test at an inconvenient moment, for example, the physician in charge (at the top of this specific hierarchy) simply decided to request the test a day earlier: "I think we can adjust that a little," he remarked. Similarly, recall Mr. York, who was transported for x-rays and other tests while the protocol involved already demanded "isolation measures." Medical personnel resolved the conflict, and thus kept the protocol functioning: the tests were a key part of the protocol. Keeping the tool on track, here, necessitated violating it at the same time.[31]

Second, tools inevitably end up in situations where it just is not clear whether a rule applies or not. This is often the case when it has to be decided whether a patient is eligible or suitable for a tool. Most oncology protocols, for example, work with staging tables to determine the extent of disease. For lymphomas, the "Ann Arbor staging" has as stage II: "involvement of two or more lymph node regions on the same side of the diaphragm or localized involvement of an extralymphatic organ or site and of one or more lymph node regions on the same side of the diaphragm."[32] In a lymphoma patient presenting with a "firm infiltrate" in the skin above the chest bone, the discussion arose whether this was a "localized" involvement of another organ (the skin), or whether this indicated "diffuse" involvement. In the latter case, the lymphoma would have to be called stage IV ("a worse case"), and different treatment protocols would apply. What "localized" or "diffuse" means in these instances has to be renegotiated in the light of the specifics of the situation at hand.[33]

Similarly, when the coronary care unit would be full, ACORN's rules would suddenly become highly problematic because the context of their use would have been altered. Should patients then be sent to other hospitals? Or would the general intensive-care unit or even the ordinary cardiology department be sufficient if this need arose? How should ACORN's output be followed up on in this case?[34] The next fragment also illustrates how a written rule can suddenly become highly problematic.

At a Tuesday patient meeting, Bear presented Mr. Pierce, who had had a bone-marrow transplantation treatment for a "relapsed Hodgkin's." "He has had a rapid recovery after having his graft back [his frozen marrow cells, reinfused three weeks ago]. Last Friday he had more than 500 granulos [the number of granulocytes, a type of white blood cell, is used as an indication for the recovery of the immune system; the patient is getting "out of the dip"]. So we stopped the isolation procedures. Now, however, the count is way down again. I don't have an explanation for that. There is no fever. He had an [infection of the scrotum], but that's improving. I stopped all antibiotics yesterday. But his granulos yesterday were around 250, so I'm a little bit scared."

"More than 500 granulos," I had learned, was a rule used to end the isolation procedures. When this level was reached, the "door" would "open," as they said, *since it meant that one could now freely enter and leave (including the patient). The time of the "closure" of the door, before the "dip," was simply set at a fixed time relative to the start of chemotherapy. I thus asked whether they would now close the door again; whether they would restart the isolation procedures.*

"I was hoping nobody would ask me that," Bear said, and smiled. "Don't you have anything else to do?" Everyone laughed, but the bacteriologist, present also, was nodding approvingly; he agreed with the suggestion. "You propose to start isolation again...." Bear looked at me. "Give me reasons...." An intern joined in: "Well, there is an increased infection risk again, isn't there? I mean, that's the point of this rule in the first place?" John, the resident, started nodding too: "that's logical, that's logical." "But it's an arbitrary limit, right?" the bacteriologist asked. "Yes," the intern replied, "but it's the rule we work with here." Bear joined in again: "We just use the granulos to open the door, not to close it." Everyone laughed again. He continued: "Maybe we can give him GMCSF [a drug stimulating blood-cell recovery]." The discussion shifted to whether or not that would be of help. They decided to start precautionary antibiotic treatment again, but took no other action. "Let's wait for today's results," was John's final remark, before they went on to the next patient.

Pierce's blood cells slowly rose again, but remained beneath the "500 granulos" level for at least two additional days. Nevertheless, the isolation measures were not restarted.

What the "500 granulos" rule means now that the granulocytes first have opened the door and now start to act unexpectedly is an open question. As the fragment shows, Bear seems inclined to keep the door open. To him, starting precautionary antibiotics is "good enough" for now; closing the door would imply reoccupying the special isolation room, both blocking the treatment of other patients and potentially upsetting Pierce. No explicit "decision" is made on this matter: as often happens, further action is postponed. John does not raise the issue again, and the door remains open.

Anticipating futures
The tools perform tasks by going through a fixed sequence of intermediate steps, demanding the completion of one step before the next can be made. In medical work, on the other hand, next steps to take are often anticipated, and current activities are often adjusted accordingly. In the account about the lung cancer patient receiving a CT scan, the protocol called for a roentgenographic evaluation of the effects of the therapy only after the whole treatment schedule was finished. The physician, however,

anticipated a failure of the tumor to react. He ordered the evaluation immediately—which, in the end, led to the patient being taken "off" the protocol. In another situation concerning the same protocol, the physician in charge had altered the treatment scheme: the second course of chemotherapy would now be given at "day 15" instead of "day 16." Upon my query, the physician said: "This woman seems to have long dips [i.e., her immune system recovered slowly from earlier courses of chemotherapy]. If we'd keep the second course at day 16, the third course would fall right into the dip—and we don't want that." Here, the protocol's sequencing is tinkered with since the physician anticipates problems with subsequent steps.

The following quote from an interview with de Dombal (July 16, 1993) illustrates this same phenomenon: "When the ward is in shambles, for instance, when it is overcrowded, or when the patient cannot communicate, or when a resuscitation is going on in the bed next to the patient, he or she is not entered [into the computer]." While in medical work the sequence, relative importance and priority of the different steps is continually readjusted to the situation at hand, the tools neglect the fact that in a more hectic situation, a briefer investigation will have to suffice.[35] Whereas ACORN, similarly, does not proceed without an ECG being taken, medical personnel may improvise on the spot. When the machine is occupied, or gives unintelligible output, staff members can proceed with other parts of the examination; when need be, they can already call the coronary care unit if they anticipate that the ECG will show an infarction. Anticipating tensions between the tool's requirement that all data are elicited and the practicalities of the situation at hand, staff members can decide to circumvent the tool, or adapt it to the altered circumstances.

Supporting Decision-Support Techniques: Some Consequences

Creating the conditions for the tool to work, thus, is not merely the accomplishment of the tool designers: it is also a real-time accomplishment of the personnel working with the tool. Rather than attempt to further discipline the practice, the task of filling the gaps between the tools' prerequisites and the occurrences in the ongoing work where the disciplining falls short is delegated to medical personnel. They pre-process data, repair output where the tool has gone astray, and reassess the pre-set rules whenever unforeseen situations occur. They omit items, deviate from the prescribed path, or even play upon the tool: purposefully manipulating the tool's input (the blood sugar values, in the account of the gynecology intern) to satisfy both the contingent, practical exigencies of the situation at hand and the demands of the tool. To keep the tool functioning, they tinker with organizational settings if necessary, or ensure the patient's compliance. Just what this means for the positioning of the tool in a practice, and of staff members with respect to the tool, will be explored in the final section of this chapter.

Additional Responsibilities

In one of the examples discussed, a physician translated pain "radiating to the back" into a category available on de Dombal's form: "pain in the flank." This seemingly trivial example reflects an issue that is of vital importance to an understanding of personnel's work with decision-support techniques. This physician not only had to translate the patient's complaint into a symptom the computer understood; *he simultaneously had to judge the consequences of this translation*. Staff members have to estimate the effects of their actions in the light of the tool's subsequent accumulating or coordinating performance.

Such judgments are highly consequential: mistranslating a data item might harm the patient by setting the tool on a wrong track. A symptom like "a pain radiating to the back" has diagnostic significance, according to most textbooks: both acute pancreatitis and acute cholecystitis can be accompanied by such pain.[36] The physician has to judge, then, whether by translating this complaint as "a pain in the flank" he might end up feeding this tool information that would lead it to miss a crucial diagnosis.[37] Similarly, intervening in the protocol's course of action can put the patient's well-being at stake. The nurses who were wheeling Mr. York around while the protocol demanded "isolation measures" had to judge whether this action would not lead to infections when the protocol's next step, the chemotherapy, would hit York's immune system. Likewise, doing a complex test a day earlier, at a more convenient time, might cause exactly the effects that test set out to screen to be missed. It was such potential consequences that upset the patients working with GLADYS: not being able to oversee the repercussions of the continuous translations they had to make was thoroughly disquieting. How could they know that their understanding of "heartburn" matched the machine's? Might they not, by accidentally misunderstanding a question, end up with a wrong diagnosis? For each and every repair, streamlining, or deviation, those working with the tool have to judge the interventions as to their adequacy in the light of how they might affect the subsequent diagnostic and/or therapeutic steps of the tool.

The consequences of tinkering with the tool are not restricted to the impact the tool's subsequent actions might have on the patient. Since the tool's implementation in a given practice thoroughly restructures that practice, the consequences that have to be anticipated are as heterogeneous as the practices themselves. Interventions might generate troubles for medical personnel dealing with the tool later or elsewhere. Advancing the date of a screening test in a protocol, for instance, should not lead to timing problems later in its course. Also, consider the following remarks from one of the nurses taking care of the deteriorating cancer patient mentioned above (the remarks were made just after the CT scan had been made, but before the decision came that the patient would be taken off protocol):

Her clinical condition is worsening. Her diarrhea is severe, she's got a fever, and her pain is intense. Now the protocol says that she needs to go to East Hospital [a 30-minute drive] tomorrow for radiation treatment. But I don't think we can do that to her. And it's a pain for us too: with her condition, we'll have to send along an additional nurse in the ambulance. We're short on staff here—and they might send her back anyway because in this condition she's probably not even suitable for radiation treatment. (interview with nurse Carl Simons, January 6, 1993)

The decision whether to intervene in the protocol's course, Carl makes clear, has consequences for the well-being of the patient and for the organization of a days' work on the ward: both will have to be taken into account in figuring out how to tinker with the protocol.

Indicating "pain in the flank" on a form when the patient had complained of "pain radiating to the back," in addition, might be construed later as a medical error. Similarly, manipulating a protocol's input may infuriate one's superior. Here, it is not so much the effect of the tool's further workings on the patient that has to be taken into account, or an increased workload for colleagues. Like the making of summaries discussed above, a decision-support tool can function as a document for "'making evident' . . . the rational and standard character of their actions" to which superiors and colleagues orient themselves (Whalen 1993; see also Zuboff 1988). Hence, in tinkering with a tool's input or repairing output, personnel often take this potential role into account.

In dealing with the tools, thus, personnel not only have to avoid detrimental consequences of the tool's exigencies in the situation at hand. In the articulation of the tool's demands on their subtasks in managing a patient's trajectory, personnel cannot only be concerned with the immediate impact of the articulations construed for the present situation of the patient. In addition, *they have to continually gloss their activities in terms of the repercussions this intervention will have within the formal system* (Suchman 1993b). In the words of Agre (1995), the formal representation of the staff's work practices is not merely something posited by tool builders and ascribed to staff members' activities. Rather, this representation becomes a concrete, active element in the activity itself, whose reactions to all the tinkering mentioned must be continually anticipated. Telling ACORN that no fever has occurred, since an inflamed toe could never be significant for the tool's purpose requires a judgment as to that purpose—which requires knowledge of the way the tool performs its accumulating task.

The burden of this additional responsibility becomes even more salient when one realizes that glossing one's actions in terms of a tool's logic is no easy matter. The tool's distinct subtasks—coordinating activities over time and space or accumulating arrays of data—articulate with and transform the subtasks of medical personnel. And, just as physicians have only a cursory knowledge about nursing work, and a pharmacist may perfectly perform his subtask in managing patients' trajectories without knowing much about physicians' work, medical personnel's knowledge about a tool's subtasks is also generally cursory.[38] Should the lymphoma patient with the patch of hardened skin have been included in a protocol for stage I and stage II patients? To be able to truly judge this, one should have known all the intricacies, effects, and side effects of the protocol's treatment schedules, the treatment's relation to skin involvement, the way this staging is performed in other hospitals working with this protocol, its effect on the research goals of the protocol, and so forth. And what will the consequence of translating "pain radiating to the back" to "flank pain" be? Although a physician might form a general impression, the exact way this translation will reverberate within the statistical formulas of the tool is beyond everyone but the tool's designers.

The negotiation processes of which the tool is the result complicate the task even more. Since it results from an eclectic mix of logics (among which medical personnel's expertise is only one), the tool's functioning is often incomprehensible to the personnel even if they try to acquaint themselves with it. In a related vein, Weizenbaum (1976) has pointed out the fundamental opacity of modern computer programs. Since these programs are the end result of a protracted, collaborative process of interactively adding fragment to fragment, there is no one point from which "oversight" of the whole program is realistically possible. Nobody "knows," to the core, what such programs do: even the designers of ACORN, with its relatively simple rule base, are sometimes surprised by the systems' behavior (interviews, Wyatt and Emerson).[39]

Once part and parcel of a practice, thus, formal decision tools create both a need for ongoing repair and prevention and a pressing necessity to continually oversee the repercussions of this work. Since this is only partially possible, a continual, tentative "groping in the dark" as to the consequences of these deviations and repairs is inevitable. While accumulating and coordinating tasks are redelegated from personnel to tools, then, this does not lead to an unequivocal simplifying or "streamlining" of personnel's work, or to an elimination of "information overload." The situation is much more complex. While tasks are redelegated from personnel, the tool at the same time *adds to* their subtasks by generating additional, intricate responsibilities. The fact that medical personnel have to continually achieve the tool's prerequisites nullifies any hope for a simple, unidirectional redistribution of tasks from burdened staff members to decision-support tools.

Localization Revisited

Faking data, disobeying a tool's output, or improvising to grasp the inner logics of a tool: these activities seem at odds with what the tool builders were trying to achieve. In a way, the tools seem to be saved by the same qualities that they set out to erase: the personnel's tinkering would certainly not appear to be "scientific" from the point of view of the tools' discourses. This would, rather, look more like the portrayed characteristics of the irrational, cognitively overloaded physician causing medical practice's problems. As Gasser (1986) phrases it, however, "far from acting irrationally, the informal practical actions of participants actually make systems *more* usable locally."

Seen from the perspective of medical personnel, the tool appears as an additional element setting constraints on their subtasks—an element that is tinkered with for purposes at hand. In oncology wards, for example, research protocols often figure as treatment options to choose from; rather than judge the "fitness" of a patient for a protocol, then, physicians are considering the fitness of the protocol for the patient (Timmermans and Berg 1996). Within limits, a tool's actions can be anticipated by staff members: the ways they articulate with the subtasks of nurses or physicians have become partially "transparent" to the latter. Local theories come into being about how to deal with lymphoma patients with skin lesions, or how to get "radiating pain" on the form, or how to cope with a full coronary care unit. Flexible interpretations of the tool's output arise: when the blood-cell counts seem to recover well, a next chemotherapy course is started even if the protocol's lower limit is not exactly reached yet. These local theories are tied to local notions about the (non)functioning of the tool, about when it should be used and when it should not—and, since nurses and physicians' subtasks articulate with the tools' task differently, these theories will differ between them. Such "custom fitting" occurs whenever a tool becomes embedded in concrete medical practices: the development of "procedural dialects" is ubiquitous because of the incessant stream of conflicts between the multiple and contradicting exigencies that are at play (Jordan and Lynch 1992).[40]

The tool does not become the ideal-typed, central decision maker or rational pathway it was proposed to be, requiring personnel only to feed data and maybe fill in some details. The tool does not do away with the *ad hoc*, heterogeneous work of managing patients' trajectories. This is not a deplorable and preventable outcome of the "corrupting" processes of getting a tool to work: it is the only way for the tools to work in the first place. Delegating the task of producing the tool's demands in real time to medical personnel *requires* leaving them the leeway to digress from the tool's prescribed steps, to skip or skew input, or to sometimes just avoid the tool completely (cf. Lipscombe 1989; Star and Griesemer 1989). It requires allowing medical personnel to adjust the tool to their ongoing work. It requires that the tools become part and parcel of local work routines. It requires, thus, a further *localization* of the tool: a moving away from its ideal-typed universality and uniformity.[41]

A Distributed Locus of Control: The Transformation of Medical Work
I have argued above that the "control" of the patient's trajectory lies neither with the physicians nor with the tools: the overall trajectory emerges out of an intricate interlocking of subtasks, of which the physicians and tools perform some. How does the above analysis fit into this picture? Does the fact that staff members may adjust the tool to other arising exigencies imply that the locus of control is shifted back to them, as some critics would argue? The answer is no. First, we saw that the idea of "total transparency" is an illusion. The personnel's grasp of the tool's functioning is incomplete, which implies that a simple delegation of control back to staff members is never achieved.

More fundamentally, arguing to shift control back to medical personnel ignores the tool's position as an active agent—a coordinator or an accumulator—within a medical practice. We cannot conclude from the tool's inability to stand on its own that the work of personnel does or should unequivocally regain command. We cannot be content arguing that the formal representation is an impoverished version of "what really goes on," since *the formal representation has become a highly consequential part of "what really goes on."*

Formal tools, moreover, can fulfill these functions *because* they are "impoverished" versions of "what really goes on." Only through this "detachedness" can they hope to be transported from the situation in which they are produced in the first place; only through this packaging can they *be* in locales different in space and time and perform their subtasks. Only through simplification (reduction of complexity, deletion of details) can workable tools emerge. They fix a history of negotiations and allow (or require) medical personnel to work from there. It is exactly through this fixed, black-box character that they allow personnel to draw upon accumulated past work in the present (Wood 1992).

Through personnel's work to keep the formalism functioning, thus, the tools create a new world. In becoming a participant in the practice it represents, the representation transforms this practice: new decision criteria come into play, more elaborate treatment schemes can be dealt with, more actions spread out in time and space are coordinated, and so forth. In reclaiming some of the control, staff members make the tool's controlling them—transforming their world—possible.

Who, now, is controlling what? Or what is controlling whom? We remain in a situation where control is distributed. There is no one person or thing in control. There is, rather, a hybrid, a powerful amalgam of heterogeneous elements, through which control is dispersed in intricate ways: dispersed over the entities performing the subtasks, and further complicated by the personnel's task of ensuring the tool's functioning in real time. Only through the protocol can medical personnel achieve the orderly administration of oncological therapies; only through the personnel's work to achieve the formalism's demands, similarly, does the protocol function. There is no one person or thing to be held "responsible": only through ACORN can nurses admit patients to the coronary care unit, but only through the nurses' producing input and reacting to output can the tool fulfill this role.

This balance is not easily shifted. Granting situational control to ACORN leads to chaos at best—to queues in front of the coronary care unit when the latter is full, to patients' torturing themselves to answer all the required questions in the required format' and to erroneous decisions. Arguing, on the other hand, that situational control belongs (or should belong) "wholly" to the physicians or nurses overlooks the fact that practices *are* changed by the tools, and that these changes occur only through the delegation (and subsequent transformation and loss of transparency) of tasks from staff members to the tool.

6
Producing Tools and Practices

In 1970, William Schwartz predicted in the *New England Journal of Medicine* that computer-based decision tools would "fundamentally alter the role of the physician" and would "profoundly change the nature of medical manpower recruitment and medical education." Schwartz foresaw that "in the not too distant future the physician and the computer will engage in frequent dialogue, the computer continuously taking note of history, physical findings, laboratory data, and the like, alerting the physician to the most probable diagnoses and suggesting the appropriate, safest course of action."

Seventeen years later, Schwartz et al. (1987) had to conclude that "the revolution has not occurred": computer-based decision tools in routine use were surprisingly rare. Many Bayesian or rule-based systems had been made, but few had been implemented successfully. A 1993 list of "Artificial Intelligence Systems in Routine Clinical Use" mentions only 18 systems.[1] Most of these, moreover, were used in only a small number of practices, and many had never left their site of origin. MYCIN, for example, has attracted much attention in the medical press, but it has never been put to practical use; nor has INTERNIST, another famous rule-based system, which was designed to handle the broad domain of internal medicine.[2] One of the systems included in the 1993 list is ACORN, which, in fact, is currently leading a dormant existence. Emerson has taken the system home: among others, nurses were not using the system frequently. "The impact of information technology in the diagnostic aspects of medical decision making [has been] limited," Barnett et al. (1987) argue; they mention de Dombal's system as the only exception. As the computer scientist Van der Lei put it (personal communication), most systems are "dead."

Decision-analytic techniques have met with a similar fate. Here too, the expectations were high:

In the 1970s, there was optimism that the application of decision analysis to clinical problems would rapidly disseminate. Some envisioned computer terminals in every clinician's office that could tap data banks filled with estimates of key variables used in decision-analysis models, and decision trees were envisioned that could be applied to the care of individual patients. Even television commercials depicted clinicians in white coats moving from the bedside to the computer to take advantage of this electronic network of information. (Detsky 1987)

Unfortunately, Detsky continues, "this optimistic vision has yet to be realized." As a tool for making decisions on individual patients, decision analysis is hardly ever used. In a nevertheless enthusiastic appraisal of the technical state of the art, Kassirer et al. (1987) mourn that "house officers rarely use [decision analysis] or even request it, even for tough clinical problems; practicing physicians more or less ignore it; and many medical educators are frustrated because their efforts to teach it to students have been wasted by the lack of exposure in the postgraduate years."[3]

A 1990 editorial in the *Journal of the American Medical Association* notes that there is still an urgent need for ways to "free decision theory from its restrictive academic base and provide a place for it in clinical reality" (Flanagin and Lundberg 1990). When I started looking for clinical applications of decision analysis in 1992, the situation seemed gloomy: several informants pointed at the Clinical Decision Making Unit in Boston as the *only* place in the world where clinical decision analysis, in its original form, was being used.[4]

The story of protocols is more complex. Research protocols are ubiquitous in many specialized fields (especially oncology). However, the protocols themselves are rarely studied: they seem to be a mundane, taken-for-granted part of the huge enterprise of clinical research.[5] This relative neglect, taken together with their ubiquitous presence, was one of the reasons I looked at some of these protocols in more detail.

The impact of national-consensus-based guidelines has remained doubtful. Despite the fact that an industry has grown up around the creation and diffusion of such protocols, several studies have shown that physicians are often not even aware of their existence—and those who are aware of them often report disregarding them (Kanouse et al. 1989; Van Dijk and Bomhof 1991). Even when physicians report having changed their practices in accordance with the guidelines, actual practice patterns often show that they have not. In an evaluation of the effects of four NIH consensus development guidelines, for example, Kosecoff et al. (1987) concluded that "taken as a whole, these four consensus conferences had no effect on physicians' hospital practice."[6] "Those who pay for and review medical care," other authors conclude, "bemoan the lack of impact on clinical behavior of practice guidelines" (Lomas 1993; see also Eisenberg 1986).

Whereas much effort has been spent on evaluating the effects of national consensus guidelines, considerably less is known about local protocols (ranging from regional, interdisciplinary protocols prescribing the approach to suspect skin lesions to ward-specific protocols on the proper way to administer a certain drug). Clearly, and in contrast with computer-based tools and clinical decision analysis, local protocols are a common feature of current medical practices. Owing to the differences in their "goals, methods, formats, and degrees of precision," however, their overall impact is impossible to estimate (Pearson 1992; chapter 4 above). Moreover, they vary widely in the way they are being adhered to: some protocols are followed by the rule, while others are "available in case somebody does not know what to do," as a surgical resident expressed it to me. All in all, studies of protocols in use show a great deal of deviation from the prescribed path. In their study of a sore throat protocol, Grimm et al. (1975) found that physicians often prescribed different types of antibiotics, or did not record data perceived to be "crucial to sound clinical decisions."[7] Similarly, Wachtel et al. (1986) studied protocols that tried to reduce health-care costs by limiting unnecessary laboratory testing, drug use, and length of stay. They note that, although some effect was measured, "the average utilization of services in the intervention group was higher than proposed in the protocols." Furthermore, "protocols were frequently broken," and "participating physicians varied in their ability or willingness to enforce strictly the use of the protocols with the housestaff" (ibid.).

Analyzing the Lack of Success: Of Right Things and Right Places

Why has the computer-aids revolution failed to occur? What explains the lack of success of so many decision-support techniques? Both the tools' advocates and their critics pose these questions, but they answer them in wholly different ways. After briefly depicting their positions, I will argue that both positions are problematic. Recapitulating the main themes of this book, I will contend that the ways the critics and the advocates frame their analyses lead them to miss crucial opportunities to understand and intervene in the processes described. The issues elaborated here are not limited to decision-support techniques or to medical practices: they come into play, I will argue, wherever "rationalizing" technologies are to be implemented in work practices.

Obstacles to the Tools: The Advocates' Analysis

According to its advocates, we have seen, decision analysis is a universal technology: in principle, the technology can be used everywhere, by anyone, on any medical problem.[8] Why, then, has the "permeation of decision analysis into routine clinical practice" been "sluggish" (Kassirer et al. 1987)? As the following quote suggests, there is no single answer.

> Several factors appear to be at work: a large number of highly competent teachers of the method are not available; few practitioners are skilled in its clinical use; computer programs to support decision analysis that are as simple and convenient to use as word-processing programs are not available; time required to complete a satisfactory analysis is costly; and efforts to enhance the "marketability" of the method by targeting specific audiences have not evolved. (ibid.)

What is notable about this list is that most of these issues concern the *setting* of which the tool has to become a part: too few teachers, too little time, and too little money are available. Likewise, other authors argue that psychological obstacles hinder physicians' enthusiasm, including "the traditional reverence for intuitive judgments" (Detsky 1987) and the fact that outcomes of decision analysis can "feel wrong" because the analysis does not share the physician's biases (Balla et al. 1989). And, it is added, more attention should be paid to generating the data needed for decision analysis—data which are often not available (Cebul 1984). These are the obstacles that prohibit widespread deployment of the technique. The techniques themselves are not to blame: the places where the tools are to work are at fault. Although some problems still remain with the measurement of utility, for example, to the advocates these are not fundamental. Balla et al. (1989) argue that "improvements are needed," but this need does not invalidate the current potential of the tool (Holmes-Rovner 1992). Through educating doctors, obtaining more funds, and smart marketing (through, e.g., targeting the message to "role-model" physicians), clinical decision analysis might still become a success.[9]

When one looks at computer-based decision tools, a similar picture emerges. The overall tone is somewhat more subdued; most articles combine a critique of the stubbornness of their tool's setting with a mild self-critique. Again, the tool is universal in principle: current limitations result from obstacles that, with some additional effort, can be overcome.[10] Edward Shortliffe, one of MYCIN's creators, argues that perhaps nothing "accounts more fully for the impracticality of early clinical decision tools than their failure to deal adequately with logistical, mechanical, and psychological aspects of system use"—the so-called "human factors issues"

(Shortliffe 1987). These, Shortliffe states, emphasizing both the tool's immaturity and the setting's shortcomings, are problems of both design and "resolution of inadequacies at the institutional level." Programs should not be too slow, they should not require lengthy typing, and they should use graphical displays to optimize the user-system interface. At the same time, the compatibility of departmental computer systems in hospitals leaves much to be desired, and physicians should start to accept that they need decision-support tools (ibid.; Buchanan and Shortliffe 1984).

De Dombal (1989) similarly spreads the blame. He criticizes the poor "doctor-computer interface" which limits the tools' efficiency in realtime use. But he also immediately accuses the practices in which the tools are supposed to work. Even a major improvement in interface design would not help much, de Dombal argues, because the progress of decision-support systems is obstructed by the often "totally unstructured" state of medical knowledge: "In many areas there is no agreement about definitions of symptoms and signs, or what symptoms are to be collected—or even what disease categories there are, and how they are defined." Until these problems are addressed by the medical profession, "all the computer scientist [can] do is wait." De Dombal is harsh in his judgment on the absence of initiatives to address this conceptual chaos: "In part this waiting is due to simple lack of coordination and bureaucratic inertia. . . . This lack of coordination makes it impossible to devise systems which will have wide acceptance; and is doubly surprising for often it is coupled with a plea from the relevant bureaucracy for more 'efficiency in medicine.'" Notwithstanding these obstacles, de Dombal's analysis also ends on an optimistic tone. The challenges can be overcome, and "it seems only a matter of time before the use of decision-support systems becomes more widespread" (ibid.).[11]

The small impact of national-consensus-based guidelines is likewise accounted for. In their evaluation of the National Institute of Health's consensus development program, Kanouse et al. (1989) state that some guidelines suffered from excessive vagueness and "left important concepts undefined"—like how "high risk patients" are defined: "We found that we often needed expert medical consultants to help us interpret what the panel's recommendation would mean in clinical practice, and that the interpretation we arrived at was sometimes not the only one that might reasonably be made."[12] Here again, however, the disappointing effects were due mainly to the several "administrative, educational, patient-centered, or economic barriers" to the implementation of guidelines (Lomas et al. 1989). Through educational channels, popular press coverage, publications in major journals, and direct mailings, the guidelines should be much more widely distributed, Kanouse et al. argue (1989). Simultaneously, additional strategies are necessary: to get physicians to change their behavior in accordance to the guideline, Lomas (1990) argues, "peer pressure, marketing of information and exploitation of informal communication channels" are indispensable.[13]

Tools That Cannot Work: The Critics' Analysis

To their advocates, the lack of diffusion of decision tools is due primarily to resistances in current medical practice. In principle, the decision tool has a universal reach. It is just blocked, here and there, by insufficient funds, stubborn physicians, and bureaucratic obstinacy.[14] The technologies themselves have their shortcomings, but these can be remedied by advancing on the current path of development.

The tools' critics turn this argument upside down. They chastise the tools for fundamentally misunderstanding the world in which they are to function. It is not a matter of "fixing" the settings' shortcomings: according to many philosophers, sociologists, and anthropologists, the tools "are all on the shelf" (Forsythe 1992) because they are *essentially* misconceived. According to these critics, formal models fail to capture human expertise: decision-support tools attempt the impossible. Intuitive, involved, skilled "knowing how" can never be replaced by detached, objective "knowing that" (Dreyfus and Dreyfus 1986); human beings can "act" (i.e., they have intentions, which can be understood by other social beings), whereas machines can only "behave" (i.e., machines' activities can be exhaustively described by their physical coordinates or by a set of rules) (Collins 1990). Where the advocates emphasize the (in principle) universal character of computer-based decision techniques, the critics point out that these tools can operate only in very specific, confined areas. A tool can be "successful" only when the task it is to perform does not require skilled "knowing how"—when the tool "takes over from us" actions that we already do "through machine-like actions" (ibid.). Moreover, one can try to "capture" social practices in formalisms, but the *ceteris paribus* condition ensures that every such codification is necessarily tied to the specific context in which it was generated. (See pp. 5 and 6 above.) The "typical" abdominal pain patient on whom de Dombal's tool is used is only "typical" of the specific population that visited the emergency ward of that hospital, in that particular period—since that is the population on which the tool's formulas are based. The only "universal" element in the discussion, according to these critics, is that "expertise" is always context-bound. No matter how thoroughly spelled out, knowledge is linked to specific practices, and is therefore always local.

The Production of Working Tools: Reanalyzing "Success" and "Failure"

The advocates have a hard time explaining just why computer-based tools are so good if so few are in actual use. Dreyfus (1992) sneers that these advocates sound like people who, having climbed into a tree, say that the moon is near. On the other hand, research protocols are used in many wards all over the Western world. From a critic's point of view, these highly detailed formalisms should not work either—but they do. On several crucial points of critique, research protocols are no different from computer-based knowledge systems or decision analysis: the basic, formal structure of the tools is equivalent (Reggia and Tuhrim 1985). Research protocols attempt to replace the contingent flow of reactive activities with clear-cut, inter-institutionally comparable, standardized actions. They are explicitly intended to be in widespread use, in many hospitals at the same time. Moreover, they are to take the locus of control out of the staff's hands, and to determine by means of a set of explicit rules what should happen in a certain situation.

How, then, is it possible that research protocols have become so widespread? Are the advocates right? Are the critics missing the point? If the latter, how can we understand the differences in "success" between tools? Or are both analyses on a wrong track?

The critics and the advocates seem to phrase their analyses in contrasting terms. Whereas the advocates blame medical practices for resisting the tool, the critics argue that the problem lies with the tools' misconception of the nature of these settings. Where the former stress the tools' universal problem-solving power, the latter see no truly universal feature of these technologies but their impossibility. Underlying this opposition, however, lie some strong assumptions shared by both critics and advocates. The structure of their analyses, the central questions they ask, follow a similar pattern. In attempting to explain the success or failure of decision-support techniques, both critics and advocates debate the *nature* of tool and practice. This mode of arguing renders "tool" and "practice" into two fixed categories, with fundamental properties which are to be listed and compared. Both critics and advocates want to judge the actions of formal tools in comparison with those of human experts, and query whether the tool is an adequate *representation* of the structure of medical practice, of the physician's decision making, or of an optimal doctor's performance. Both analyses, then, *contrast* tool and practice by reifying them as separate categories, comparing them, and distributing blame for failure or praise for success.

Critics, for example, state that the computer cannot but fail as a decision-support technique since it cannot operate, behave, or think like a physician—it lacks the agency, the "intentions," the "tacit knowledge" humans have. For Dreyfus and Collins, human practice is endowed with fundamental qualities that machines simply do not have, and thus every attempt to capture skilled performance is doomed to falter. Human practice is pitted against machine action, and the argument attempts to demonstrate the ontological and epistemological gulf that forever separates humans from techniques. As Robinson (1994) nicely phrases it, critics "put intimacy and abstraction at opposite ends of the same continuum": in criticizing the formalisms, they emphasize the poverty of these representations with respect to the richness of the actions represented. Whether we are dealing with expert systems, or with statistical formulas that would represent the physician's mental processes, or with protocols that would represent a sequence of practical activities, critics expose the *a priori* failure of a map to capture the intricacies of the domain the map represents.

Advocates also debate the nature of human practice and machine action. Whereas critics argue that these categories are separated by a deep gap, advocates argue that they share crucial similarities. According to some advocates, for example, both tools and medical personnel belong to the species of information-processing agents; according to others, both tools and practices are structured according to a scientific, step-by-step approach to clinical problems. Here too the issue is framed in terms of just how much these categories are alike, and just how well the tools represent the decision making or actions they embody. For example, advocates often attempt to evaluate computer-based tools by comparing the computer to a human expert on the basis of how each deals with a similar set of (paper) cases.[15] Instead of putting intimacy and abstraction at opposite ends, advocates would rather argue that the former can be rendered into the latter without danger—that the tools can adequately represent a physician's decision making or a practice's management strategies.[16] Here, tool and practice are contrasted not to demonstrate their difference but to argue their similarity. Since the reason for failure, then, cannot be found in fundamental properties of either tool or practice, it must lie elsewhere: in current imperfections of the tools, and in contingent psychological and socio-economic barriers in the practices.

However, contrasting human practices and tools to compare the two, or to distribute blame and praise, is not a fruitful way to frame the debate. Debating success or failure by pointing to fundamental qualities of human practices and of tools overlooks that these qualities themselves are not pre-given. By basing their critique on the unique properties of human practices, or, respectively, by positing the existence of the universal category of information-processing species, critics and advocates assume a foundation from which to argue difference (the critics) or similarity (the advocates).[17] Such foundationalism, however, overlooks that what is taken as foundation, as "basic category," is itself already a product of the historical intertwining of humans and tools. What is our understanding of "humanness" other than a conglomerate in which a whole range of metaphors (the steam engine, the telephone switchboard, and, yes, formal technologies) play roles? How can we hope to disentangle "human practices" from the technologies that structure them, the artifacts upon which they feed? And do we not describe, develop, and work with computers in terms directly derived from human "properties" and activities such as "memory," "writing," "processing," "messages," "hierarchies," "commands," and indeed "intelligence"?[18] Equivalence between tool and practice is (or is not) achieved in and through the transformation of discourses and the disciplining of practices: in and through the setting up of psychological experiments (including the use of paper cases to compare man and machine), the explosion of computing metaphors for the "mental," and the transformation of diagnostic and therapeutic tasks. The *nature* of, say, "medical practice," "expert systems," or "physicians" is the outcome, the *effect* of these intertwined transformations (Law 1994).

In the early 1960s, statistical techniques similar to those used for the first statistical diagnostic tools were employed in an attempt to ameliorate the disordered state of medical taxonomy: to eliminate the "logical fallacies which abound in the subjective, orthodox methods" of disease classification (Sokal 1964; Jacquez 1964; Overall 1972). If these approaches had had more impact, if medical taxonomy had been restructured in this way, the statistical tools would have transformed medical practices more to their liking—and they could, then, have been more "successful." Also, through the proliferation of biophysical technologies, monitoring instruments, and so forth, qualitative data are increasingly being replaced by "digital" data.[19] Here again, practices change and become more amenable to formal tools. On the other hand, more and more builders of computer-based tools are adjusting their tools to the resistances experienced in earlier attempts. Tools are built that function more in the background (as "checklists"). Tools are integrated more into other, larger information technology products, and so build upon infrastructure already in place.[20] In the processes that have led to these current states of affairs, then, both practices and tools have been transformed.

If this is so, the very question *why* a tool is said to be successful or unsuccessful becomes problematic. "What accounts for failure or success?" tends to be read as a general question requiring a general answer and resulting in a list of "factors" or "aspects" which, by their general nature, are easily read as qualities of a tool and/or a medical practice. For example, the widespread use of the research protocol can be "explained" by referring to its rational, scientific character, or the "lack of success" of expert systems can be denoted as a "matter of time" by pointing at the fundamental equality of the nature of the tool and the nature of medical thinking. Alternatively, the research protocol's ubiquitousness can be accounted for by pointing to the research careers that are tied to the use of these protocols, to the existence of regional coordinating offices that aid in the proper application of protocols and process and check collected data, and to the many international medical research funds that are centered around the development and implementation of research protocols.

To turn "successfulness" or "unsuccessfulness" into a property of either tool or practice, however, and to explain "success" or "failure" by contrasting the two, is to isolate the tool from its contingent history. It is not *because* the research protocol is so rational that it has become so widespread: the specific rationality and shape of decision techniques emerged together with the development of these tools and the specific configurations of their use.[21] Likewise, a tool's setting cannot explain its "success." To the contrary, I would conjecture that protocols have played a significant role in shaping both the content and the context of current oncology: it is probably *because* the emerging protocol has been an ubiquitous element of this field since its early days that its diagnostic categories and therapeutic interventions are as complex and elaborate as they are now.[22] To argue that the expert system is a failure because medical practices resist it is to overlook how central the cognitivist perspective has been to the redrawing of the nature and the flaws of medical practice. The questions of critics and advocates pry apart tool and practice instead of relating them; they freeze the two into an artificial "snapshot" opposition instead of following their intertwined trajectories. There is no meaningful way to talk about a tool without at the same time speaking of the practice with which it coevolved: the research protocol is part and parcel of the network that is "current oncology." One cannot cut away either one or the other without losing sight of the constituent features of their current "essences." The "properties" of both tools and practices are in constant motion. And when nothing has remained the same, when tools and practices have become so thoroughly intertwined, blame or praise for a tool's fate cannot be distributed unequivocally. A story of heroes or villains has lost its ground. "The tool" or "the practice" can no longer serve as an explanatory category: an understanding of the current situation requires an understanding of the way these categories are intimately involved in each other's production.

Changes and Convergences

Nothing is, by itself, either reducible or irreducible to anything else. There is no equivalence without the work of making equivalent.
—Bruno Latour (1988)

In this book I have focused on how specific tools and practices have evolved together. Whereas critics and proponents primarily center on the tools as *product*, I have looked at the *process* of the construction and implementation of these tools, and at the discourses that came into being with them.[23] The tools and the worlds in which they become embedded thoroughly transform each other—and these mutual transformations are key to an understanding of their (non-)functioning. A working tool, I argue, is the outcome of these mutual transformations: of the *convergence* of tools and settings into a network in which heterogeneous elements are interconnected and transformed (Bowker 1994b).

There are many interrelated layers to this process. I focused, first, on *medical literature*, describing how with the coming of the tools new discourses came into being. These new discourses redefined what medical practice is, what its problems are, and how these problems should be addressed—and so in a way that *made* the tools "the right ones for the job" (Clarke and Fujimura 1992). By redescribing medical practice in the image of the tools, the discourses achieved a smooth fit between tool and practice. In discussions on medical practice and its problems, the tools became obligatory points of passage for many attempts to rationalize medical work.[24]

In the *system builders'* construction and implementation of individual tools, medical practices and tools are made to fit each other. Practices are disciplined to ensure appropriate input and adherence to output. Concurrently, the technique is transformed: in inscribing locality into the core of these tools, they are redescribed in the image of the practices.

In the work of *medical personnel*, finally, the same mutual transformations return. Where the disciplining of a practice falls short, medical personnel can achieve the formal tools' prerequisites in their work. In doing so, they help the tool transform the practice, extend and/or complexify the local network—and the tool is further localized.

All these practices intermesh: I have been describing a complex network of medical personnel, blood cells, journals, decision techniques, system builders, patients, and so forth. This one network can at the same time be seen as the interlocking of several networks, all exerting their own sometimes-conflicting demands. Physicians and nurses are monitoring and shaping their patients' trajectories *and* they are themselves an element in the network built by the tool builders; system builders and physicians are influenced by journal articles *and* they themselves participate in the very activity described in these texts. Shifting the point of entry from medical literature to the work of system builders to the work of medical personnel, then, was a means to get a grasp on the different layers that are at stake in the development of decision-support tools, and on their interrelations. We can never hope to comprehend these multilayer processes if we attempt to stick to one point of entry for our analysis: the network is not set in motion by one central actor. To get a hold on this multidimensionality, it is a prerequisite to shift positions throughout the network—to attempt to grasp the way the different layers can be tightly interconnected yet have distinguishable dynamics of change.

"Convergence," then, points at the way technique and setting mutually transform each other to each other's image—whether the focus is on medical discourses, on the work of system builders, or on the work of medical personnel. It points, also, to the way these various activities and occurrences (including those I have mentioned only tangentially) mesh together into niches for the coevolving decision-support techniques. Only when new discourses have made tool builders' interventions possible, only when medical personnel in real-time maintain the niche tool builders have construed, only when the tool builders' attempts interlock with coinciding practices of standardization—only then can a working tool evolve.

"Convergence," also, points at the simultaneous and similar transformations of "social" and "natural" worlds (Bowker 1994b). The work of medical personnel is rewritten in the light of the tool—and vice versa. Similarly, blood tests have to fit the tool's requirements as much as the nurses' behavior. The thoroughly heterogeneous practices are disciplined through equally heterogeneous means: forms, training, bureaucratic hierarchies, pre-packaged medication, and so forth. Localization, simultaneously, is inevitable, since both "social" and "natural" elements can elude control.

And "convergence" also points at the fact that characteristics of "tool" and "practice" are not pre-given but rather emerge in and through the development and intertwining of the networks.[25] "Universality," for example, is not a static attribute whose actuality or absurdity is to be debated, as the tools' advocates and critics tend to do. It is not so much a characteristic of the tool as a possible *consequence* of extending the tool's reach through disciplining a whole array of local practices, and achieving the formalism's prerequisites. Universality is not some ethereal quality of a superior technology, but an emerging feature of the networks of which that technology has become a part. It is a *dynamic* characteristic which, in the fighting of the different types of localization, is continually in the process of becoming—and continually in the process of slipping away.

This latter feature immediately also points to some caveats that should be made explicit. "Convergence" should *not* be taken to indicate a gradual development toward one fixed end—toward some ideal-typed horizon where the networks have become stable, "irreversible," and all-inclusive (Callon 1991). Convergence is and remains a process, forever incomplete, unending, and non-unitary. "Convergence" should not be equated with homogeneity, or with the gradual emergence of one all-encompassing logic. The decision-support tools themselves, we have seen, carry different notions of what rationality is, what medical work is, what its problems are, and how these problems should be remedied. Owing to the processes of localization, implementing these tools in a wide array of medical practices only results in further fragmentation.

In addition, the processes of convergence always remain partial in the sense that we remain confronted with many loose ends. It would, for example, be wrong to assume that the discourses described in the first two chapters fully overlap and account for the specific actions of tool builders described in the chapters that followed—that de Dombal's activities and intentions, for instance, fully mapped the cognitive discourse I described. Interpreted in a mechanistic way, the notion of convergence might seem to indicate the emergence of a strict chain of necessary conditions or of a complete isomorphy—yet the interplay between the discourses and the activities of tool builders is much more fluid and partial than such a depiction would suggest. Although de Dombal's tool is an excellent example of the statistical logic described in chapter 2, whether de Dombal actually underwrote the view that physicians think statistically is an open question;

although Feinstein (1973a,b) opposes the cognitivist discourses described here, he has written papers on the "analysis of diagnostic reasoning."

This multiplicity and partiality, the existence of variances, holes and gaps in the convergence story, is not merely an empirical finding. It feeds into the very survival of this network. Thus, localization can be fought by increasing disciplining efforts, but it can never be transcended: there is no breaking away from localization, no escape from impurity. The tool persists because of the existence of loose ends and different logics. Similarly, as we saw, the connections between the activities of medical personnel and those of the tool builders ended in a situation where a network could gain a fragile stability because it never fully crystallized (Singleton, forthcoming). Medical personnel had to constantly rearticulate the tool with the ongoing course of events—but this implied that they had to be left free to reappropriate the tool in ways they saw fit. The processes of convergence, as I would like to typify them here, are beset with paradoxes. The lines of the network do not spread out in some increasingly encompassing and fortified grid: they are turned onto themselves, bent in such ways that the only way the network can persist is through its looseness, its openness, and its unresolved tensions.[26]

Different Questions, Different Issues

The point is to get at how worlds are made and unmade, in order to participate in the processes, in order to foster some form of life and not others.... The point is not just to read the webs of knowledge production; the point is to reconfigure what counts as knowledge in the interest of reconstituting the generative forces of embodiment.... The point is, in short, to make a difference.
—*Donna Haraway (1994)*

Of Maps and Terrains

Instead of contrasting tool and practice, I have focused on how tools and practices have developed together and transformed each other. Instead of trying to explain success or failure by opposing two pre-fixed entities, with pre-fixed characteristics, I have gained insight into the current (non-)use of decision-support techniques by looking at how the networks in which tool and practice coevolve have converged—or not converged.

In refocusing the questions in this way, some issues come into view that are neglected by critics and advocates. To advocates, for example, the formal, "objective" nature of their tool guarantees diffusion: it is only the *resistances* that have to be accounted for and dealt with. To them, the tools are like faithful, objective maps: representations of optimal practice that lead medical personnel toward supreme performance. They are windows on a perfect world. How could they not be universally used, except for a few lingering obstacles (Latour 1987; Wood 1992)? Advocates, however, overlook the work that makes their tools feasible—the *ad hoc* management of tools' and patients' trajectories that characterizes the work of system builders and that of medical personnel. "Disciplining practices to a formalism" is not the taking away of obstacles: it is the building of the roads on which the tools can travel in the first place. "The outside world is fit for an application of the map only when all its relevant features have themselves been written and marked by beacons, landmarks, boards, arrows, street names and so on"—that is, when the landscape is *itself* rewritten in the *same* language as the map (Latour 1987). Once, through this ongoing work, a network has been made stable enough to have a formal tool function, the advocates' views retrospectively *become* true (Bowker 1994b). Once, through the coevolution of tool and practice, settings are sufficiently transformed to the tools' liking, institutions ensure uniform terminology, meetings ensure identical laboratory tests and criteria, and technologies are distributed uniformly—then a tool's "universality" is accomplished. At the same time, however, this accomplishment is dependent upon the endless work of committees to standardize tests, secretaries to fill in and check forms, and personnel to fix the breaches between their ongoing work and the tool's demands. In the advocates' view, this prior and ongoing work "disappears into the doneness": it becomes invisible (Star 1989a, 1992). The technology seems to function on its own superb and universal power, while the work of meticulously creating and repairing the social and material infrastructure that makes this functioning possible (and in which the tool itself has taken shape) disappears from sight.

Similarly, advocates overlook how tools are always *located*; how a local context and reflections of past negotiations are built into the heart of the rational tool. The advocates' claim to unqualified universality is the illusion of tools' having no history. It is the illusion of the one transparent, objective, true, optimal documentation of the world—of the "unmarked category" that represents while itself "escaping representation" (Haraway 1991). This is another reason why "success" and "failure" are problematic terms. In view of the transformations that take place in and through the emergence of new tools and new practices, it becomes clear how "success" or "failure" is always success or failure for somebody or for something in some partial way (ibid.; Star 1991). Speaking about "success" without such qualifications is itself characteristic of the view that the one rational voice is at stake—the privileged view that seems a natural common good. Every tool silences some voices and amplifies others; every tool helps to strengthen some knowledges and helps to forget others; every map has an author, a subject, and a theme (Wood 1992). Since the tools' actions can be so consequential, it is important for both participants and analysts not to delete its past—not to forget its locatedness.

Critics also tend to gloss over this issue. If formal tools are forced upon work practices, their impoverished, mechanistic functioning will transform the work into a disjointed set of impoverished, mechanistic tasks. Where the advocates applaud the tool's universal Rationality, the critics deplore the stifling effect the tool will have on the skillful and informal work routines of medical personnel. Bracing themselves against the threat of a soulless, empty formality,[27] however, critics overlook the fundamental differences between the rationalities embedded in the tools. Formality should not be opposed to content, as the tools' critics often do: formal tools carry clinical or statistical logics in more or less impure mixtures, can and will address similar problems in wholly different ways,[28] contain different assumptions about the roles of patients, and so forth. There is not one map per terrain—there can be many maps, yielding different navigations, different foci of attention, and different themes. Disciplining practices to a formalism results in as many different practices as there are tools.

Both critics and advocates remain stuck in comparisons between tool and practice. They compare the terrain with the map: does the latter adequately represent the former? Is the image objective or biased? Is the performance (as measured through, e.g., paper cases) comprehensive or impoverished? Framing the debate in this way, however, ignores the crucial matter of how the representations are used in the practices. Many critics remain stuck in stating that the tools cannot or should not function, or that they function only through human help[29]—but they pay little attention to exactly how this changes both the content and the context of personnel's work. The intriguing feature of these systems is that they alter the work that allows them to exist—but critics do not focus on the coordinating and accumulating functions of these tools. In criticizing the formal tools' poor mapping of human practices, they make the mistake of wanting to substitute informality for formality[30]—to demonstrate the powerlessness of the formal tool in the light of the sheer dexterity and skillfulness typifying human action. When critics do address actual tools in practice, they tend to demonstrate how immersing a formalism in a practice's routines results in a complete subordination of the tool—a return, as it were, to things as they were before. Grasping the intricacies of the domain and the crudeness of the map, however, does not render the latter useless. On the contrary, the map's crudeness (or, less disapproving, "selectivity") allows it to work like a map (Wood 1992): to enable one to oversee large areas at a glance or to navigate one's way across long distances. In and through the intersecting processes of achieving a tool's prerequisites, a world is created in which hybrids of techniques and personnel coordinate tasks and accumulate data in wholly new ways, across larger distances and incorporating more complexity. Thus, through the intertwining of informality and formality—through the interlocking of representation and represented—worlds shift to new planes.[31]

In addition, comparing a medical staff member's actions to the actions of a tool overlooks the fact that a tool does not "fill" the slot in the practice previously "filled" by the medical staff member—the very process of constructing and implementing a tool alters this slot. As the end result of the negotiations in the implementation of ACORN or de Dombal's tool, a practice emerged that embedded decision strategies unfathomable to physicians. Comparing the performance of a physician to that of the tool, here, ignores that physicians and accumulating tools more often than not measure different things, combine different data, or decide on different questions. In getting a tool to work, we have seen, these issues are invariably transformed. Arguing that computer-based decision aids are not able to mimic an expert physician sidesteps the fact that most implemented decision-support techniques are not mimicking physicians' actions in the first place: part and parcel of a changed practice, they do things nobody else does or can do. Similarly, arguing that diagnostic tools can make diagnosis more accurate overlooks the fact that more often than not, what "diagnosis" means is changed. Self-perceptively, Bjerregaard et al. (1976)—co-workers of de Dombal—commented on the tool's greater success in diagnosing acute appendicitis: "It must be borne in mind . . . that the clinician is taken as having diagnosed appendicitis whenever he chooses to perform an appendectomy, whereas the computer is trying to predict whether the appendix is inflamed or not."[32]

Stretching this last argument a little might make the point more clear. It is never argued whether the CT scan does its job properly, or whether humans should be left to the job. This technology has created a *new* activity: accumulating a myriad of "simple" x-rays into one composite cross-cut image. It offers a new piece of data—and the many "decisions" and negotiations incorporated in the tool[33] remain mostly opaque to the medical personnel who act upon its output. The decision-support tools discussed, I argue, are not so dissimilar from the CT scan as they might seem: they perform tasks that did not exist before, in practices that have changed in major ways. Since the practices "before" and "after" the tool's implementation are so different, and since there are no "similar tasks" to be compared, evaluating a tool's performance by comparing it to that of medical personnel misses the point.

Thus, investigating the intermeshing of formal tools and human practices and studying the transformations in decision criteria holds an important challenge. Of course, the idea that decision processes or action sequences can be faithfully represented is as chimerical as the dream of a map without a theme. Yet this conclusion does not go far enough. The vocabulary of "representations" leads one to *compare* a tool and a practice—and to overlook the active role of the tool. What matters is how a tool *in* a practice alters the work of nurses and physicians, the lives of patients, and so forth. What matters is how the production and use of the map *transforms* the terrain—where we can go now and could not before—and what is hidden from view. The complex interactions between tool and medical practice have hardly been investigated. Just how responsibilities are diffused over collectives of doctors, nurses, and protocols, or just how tools like ACORN alter the decision making strategies embedded in a practice, has received little attention. Yet the meanings of the central legal and ethical notions of responsibility and decision shift subtly when control, as I argued in chapter 5, is distributed between a tool and personnel—when "Who or what is finally in control?" has become an unanswerable question.[34] Similarly, decision criteria subtly change, affecting who is deemed "acute," who is deemed "incurable," and so forth. When one is too preoccupied with artificially comparing performances on "similar" tasks, with artificially isolating a representation from the activities in which it is composed and used,[35] and with opposing categories that evolve together and transform one another, these crucial issues are all too easily overlooked.

The notion of "logics" can fulfill an important role here. Purified notions of "human decision making" or of the blueprint of the tool are meaningless if one wants to get a grasp on the shifting decision criteria and the specific redelegations of decision power within these heterogeneous practices. In debating and analyzing logics we are no longer focusing on either physicians or tools. The theoretical unit of attention is no longer what a medical staff member decides, or what a decision-support technique does. Different logics can be discerned and discussed irrespective of in what or who these are embedded. It is not that logics are not performed and embodied in concrete practices: I am not trying to reintroduce a category that can stand apart from the networks in which it is constituted (see chapter 1 and 2). I argue, however, that these logics constitute discernible patterns that cut across the categories of "tool" and "human practice" and can be debated *as such* (Mol, forthcoming; Law 1994). When it is exactly the interaction between formal tool and human work that yields new realities, we require a vocabulary that does not require a "sorting out" of categories before it obtains the leverage to investigate. Rather than recreate dichotomies, then, and rather than set up a procedural argument that has lost all relevant anchoring points, a focus on logics affords a grasp on the *content* of the shifting decision strategies and working patterns. Going beyond the debate whether tools can be of use, a much more fruitful question arises: Which logics will yield what consequences?

What could this look like? The different logics' notions of rational medical practice and its obstacles (embedded in the tools, the texts that accompany them, the implementation-strategies that are associated with them, the way they are (supposed to be) used, the way they are materially configured) pre-structure the way in which specific practices, "problems," and "needs" are approached, conceived, and structured (Agre 1994). In conceptualizing medical practice as a scientific and mental activity, for example, the decision-focused logic narrows its focus down to the cognitive functioning of individual staff members. It overlooks the interactive and heterogeneous nature of current medical work. Doing so, for one thing, runs the risk of solving the wrong problems. One of the reasons that ACORN was not a great success was that the coronary care unit was indeed often full (interview, Jeremy Wyatt). This organizational fact caused delays in the emergency ward, and as a result patients were sent home who would otherwise have been admitted. Focused on ameliorating the limited decision-making capacities of medical personnel, ACORN had originally been intended to reduce delays and prevent erroneous decisions by improving the decision-making process. As Wyatt stated in hindsight, "we identified a non-problem" (ibid.; see also Wyatt 1991a). Organizational contingencies caused the phenomena ACORN had set out to erase—and to which it was no solution at all.[36]

The way the decision-focused logic channels attention away from the social and material organization of medical work has other consequences, too. In attributing medical practice variations to suboptimal decision-making capacities, for example, the ways intellectual capacities are embedded and constituted in concrete practices threaten to disappear from view. Conceptualizing the individual health-care worker as a central source of trouble in current medical practice distributes blame in a highly specific manner, forgoing scrutiny of insurance arrangements, of the role of pharmaceutical companies in structuring the therapeutic options available, and so forth. It was a president of the Medical Decision Making Society who stated that the contributions of "inept doctors" to medicine's failings are much smaller than the contributions of "inadequate systems" (Berwick 1988). As a modest move in this much larger debate, then, this criticism of the decision-focused logic is an argument in favor of a logic more attuned to the social and cooperative character of current medical work.

But we can also compare the logics of individual configurations of tool and practice. As a simple example, let us return to the question of the number of lymph nodes used as a criterion for inclusion in an experimental treatment regime for disseminated breast cancer (chapters 3 and 4). The axillary lymph nodes are one of the first places breast cancer cells disseminate to; the more nodes affected, the poorer the prognosis. Setting the inclusion criterion to "ten or more" follows a different logic than setting the number of nodes to "four or more": the question who should receive this potentially dangerous, intense treatment with uncertain benefits is judged differently. In the former case, the uncertainty of the benefits and the danger of the treatment are given more weight, since the treatment is reserved for the more "hopeless" cases. In the latter situation (with relatively less ill patients), the hope for a beneficial effect is given more weight, and the risk of induced morbidity or even mortality is weighed differently. Protocols in oncology, then, can contain more "prudent" or more "bold" logics—logics that give more weight to the danger of increased suffering or that prefer to fight for the last bit of hope for a "cure."

The notion of logics thus opens up new spaces for debate: it changes the terms in which practices of tools and medical personnel can be analyzed. We no longer have the advocates' opposition between the rational tool and the imperfect practice, or the critics' contrast between the domination of the sole reductionist voice and the rich multivocality of the reduced. Rather, the different logics, embodied in the different tools and practices, can now be contrasted and weighed. One rationality might be preferable to another; the politics of one formalization might be favored over the next.

Design as Critique

These remarks lead to the final issue I want to focus on: the activity of design. To advocates of decision tools, this activity is hardly worth much debate. When tool and practice are seen as essentially structured in the same way, and when the tool is depicted as simply filling an already-existing niche in a practice, the process of design is downplayed to the technical problem of construing the physical artifact. For many critics, the notion of design is also a non-issue: since the whole idea of formal tools is a misconception, their actual production is of little interest. When this *a priori* position is left behind, however, and when the work of *constructing* a niche is considered, "design" is put in a different light. When the recursive relation between tool and practice is taken seriously, it becomes clear how design is a form of critique. The transformations that occur in the processes of constructing and implementing a decision-support tool, we have seen, are not limited to the time and place of "the decision": rather, many constituting elements of the pre-existing network are thoroughly changed. "Critique," here, is not the advocates' desire for transparency and rationality, nor is it a philosopher's in-depth analysis of the tools' or a practice's assumptions. It is about politics—about making a difference in how social relations are forged, or how patients are treated. *Design as critique*, then, points to the activities that are the trick of the trade of tool builders: it is what they *do*, but what remains out of view in the writings of the advocates. It is the activity of attempting to transform a practice toward some pre-set goal in and through the production and implementation of an artifact in which this goal is inscribed. It is attempting to achieve "social change" by rewriting pre-existing relations in the durable form of a computer system or a written set of rules (Callon 1987; Latour 1993). It is a pragmatic activity: it accepts the ongoing negotiations that constitute this path, the localization of the artifact, and (thus) the impossibility of ever achieving a goal in any pure form.

Why is this relevant? A focus on design as critique, I argue, affords a way to move beyond deconstructing the tools' claims to rationality and universality toward actively trying to *transform* their development and use (Haraway 1991, 1994). When one leaves behind both medical practice and tool as the final ground from which to judge, one gives up any pre-given notion as to the potential benefit of a tool, or its proper (non-)use. The tool can no longer be perceived as the carrier of rationality, but neither can it be seen as essentially powerless—or as essentially dehumanizing. Neither tools nor practices can any longer be criticized from the comfortable position of the outside observer occupying some high ground. In such a situation, a critical position implies immersing oneself in the networks described and searching for what is or can be achieved by new interlockings of tools and practices. Here, tools have become political agents—crucial elements in the ongoing struggles to define and transform medical work and specific practices. Looking at design as critique, then, changes *and* also broadens the critic's scope of critique by going beyond the realm of proclaiming one's dissent to actually reappropriating the tools themselves.

What could this mean? What could design as critique imply from a position that does not underwrite the advocates' rhetoric yet does want to look for fruitful consequences of interlockings of tool and practice? In Suchman's (1994) terms, this would imply a shift from "viewing design as the creation of discrete devices . . . to a view of systems development as entry into the networks of working relations—including both contests and alliances—that make technical systems possible." Finding a place for a tool within "specific ecologies of devices and working practices" (ibid.) implies crossing and reconstructing the boundaries that may exist between designers and users. Rather than sustain the illusion of "design from nowhere," it implies acknowledging the locatedness of the tool and the rationalities it embeds (ibid.; see also Markussen 1994). As is the case for personnel managing patients' trajectories, builders do not plan "success"; they tinker toward it. This implies, thus, leaving behind the illusion of diffusion without transformation, and accepting that traveling through the network is a prerequisite for construing a working tool. To get a grasp on the possible instances where different networks and network-building activities either converge and mutually strengthen each other or clash and obstruct the emergence of a working tool, it is essential to acknowledge the presence of different logics and multiple positions. This need springs both from the drive to construct a tool that might actually be used and from the intimately related requirement to ask for whom or what it would then be "a success." When the breadth and depth of the occurring transformations are taken into account, it becomes clear how the politics involved are just as multilayered.[37]

This appropriation of design as critique, moreover, implies abandoning the illusion of the one single answer and considering the knowledges a tool would validate or erase. Awareness of the many options the realm of the formal leaves open can lead to the creation of decision-support techniques that embody specific medical criteria and knowledges that are held important. As Feinstein is attempting to save "soft data" from oblivion, and insurance companies might be attempting to construe protocols to minimize costs, decision-support techniques could also be construed which attempt to mitigate, for example, oncology's often "aggressive" approach in palliative treatments (i.e., when there is no hope for cure).[38]

It implies, also, leaving behind the illusion that tools could be construed that would *not* hurt somewhere—that would not discipline, or not curtail options or ranges of activity at some place in the network. This could mean, for example, an awareness that disciplining should not occur unless something is gained. In other words, the protocol's coordinating function, or an expert system's accumulating potential should lead to personnel's acquiring new competences (new skills, new responsibilities, new drugs to deal with, and so forth; see Robinson 1991b).[39] The coordinating role of the protocol, for instance, can be put to use to allow regional hospitals to handle complex treatment regimes that used to be reserved for academic centers. Arguing for and construing decision-support techniques that use their potential for "providing a wider perspective and for creating new terrains across professional and departmental boundaries" (Wagner 1994) is a crucial strategy in the attempt to intervene in the development of this class of tools.

Producing Tools and Practices

The main subject of this book has been decision-support techniques for medical work, but the issues raised here are relevant wherever "rationalizing" technologies are debated or construed for concrete work practices. Focusing on the local, heterogeneous work that is a prerequisite for rationalizing tools to function is not only important for an understanding of the current state and practical (im)possibilities of medical decision-support techniques. It is crucial in the light of the ever-present belief in the autonomous, superior functioning of formalisms—whether we are talking about the development of an electronic medical-patient record or about the construction of standardized work plans for nursing (Star 1989a; Timmermans et al., forthcoming). It is crucial to topicalize the ways rationalizing tools are located, their histories, and their different logics—whether it is the construction of an information infrastructure that is debated or the elaboration of a disease classification (Bowker and Star 1994; Hanseth et al., forthcoming). The analytical framework elaborated here shifts the unit of analysis: instead of focusing on either the tool or the work practice (or opposing them), it is their *interrelation* that is central. This implies focusing on their historical coevolution and on their interlocking in current practices, including the distribution of responsibilities, the shifting of decision criteria, and the dissolution of the very notion of "decision." The way hybrids of formal tools, people, and artifacts produce myriads of new practices that cannot simply be captured in one-line evaluations as "increasing centralized control" or "loss of worker autonomy" urgently needs investigation—whether we are dealing with automated factory production lines or with information systems supporting office work (Robinson 1994; Dodier 1995).

Global positions "for" or "against" formal tools, then, are no longer tenable. Rather, the specific locations of individual tools can now be thematized, and the illusion held by some critics that all control has to be delegated back to the users can be left behind. Just as tools without a history are a fantasy, it is a chimera to call for tools that do not require a disciplining of practices, that do not interfere in the "informal" work processes critics hold so dear.[40] Seeing that different tools can carry different logics, seeing how different tools reshape practices in different ways, opens the way to a much more fruitful strategy. Breaking away from having to either embrace formal tools or denounce them by shifting the terms of the debate creates new space, new leverage, and new potentials for intervention, comparison, preference, and maybe even choice. Design as critique, here, implies acknowledging both the dangers and the opportunities that formal tools might entail. No longer denouncing tool or practice, it means searching for ways in which such tools may become familiar yet never totally transparent, powerful yet fragile instruments of change.

Notes

Introduction

1. Many alternative names have been given to the techniques I discuss: "formal methods for decision making," "decision aids," "intellectual technologies," and so forth. Sometimes the word "algorithms" is used as a general category, but this term is more often used for a type of protocol—although conceptual variation abounds here too.

2. See Silverstein 1988 and Balla et al. 1989.

3. On clinical decision analysis see e.g. Balla et al. 1989; on expert systems see Reggia and Tuhrim 1985 and Shortliffe et al. 1979. On the "natural" place of protocols in medical practice see e.g. Margolis 1983 and Eddy 1990c. Forsythe (1993a, 1993b) and Kaplan (1987) analyze these views with regard to computer-based decision tools.

4. This group of "critics" encompasses a diverse and not necessarily unanimous set of authors. All in all, the computer-based tools have had the most attention. For philosophical critiques (often closely resembling Hubert Dreyfus's line of reasoning) see Winograd and Flores 1986, Searle 1984, and Collins 1990. For an application of Collins's ideas in the medical domain see Lipscombe 1989 and Hartland 1993b. For an anthropological perspective yielding similar conclusions see Nyce and Graves 1990, Graves and Nyce 1992, Forsythe 1992, and Forsythe 1993b. Some tool builders have become critics: see Engle 1992 and the cautious editorials of the early "advocates" Barnett (1982) and Schwartz (1987); see also chapters 3 and 6 of the present book. Similar points are made in the new field of computer-supported cooperative work; see Button 1993, Greenbaum and Kyng 1991, and Star 1989b.

5. I quote the 1992 edition throughout. The core of the argument, as summarized here, is derived from the new introduction to the 1979 edition, yet it is a rephrasing of the "main argument" made in 1972.

6. See my discussion of de Dombal's system in chapter 2.

7. Many builders of computer-based tools say that they do not want to supplant physicians; they only want to support and aid physicians in their diagnostic or

therapeutic work. However, Lipscombe (1989) argues, many systems built by these authors still offer solutions to medical problems instead of advice; they still grant the situated control of the decision-making process to the computer. For more on this see chapter 5 of the present book.

8. For some recent overviews and appraisals see Latour 1987, Bijker and Law 1992, Pickering 1992, Clarke and Fujimura 1992, Lynch 1993, and Star 1995.

9. For an introductory discussion of the notion of "formal systems" in the context of artificial intelligence see Haugeland 1985.

Chapter 1

1. The focus was not exclusively on literature from the U.S., but the journals scanned were U.S. journals. The details of the developments, therefore, should probably be seen as specific American events. However, the larger outline of the picture sketched here is one which can be seen in the Dutch and English literature as well. (For a discussion of the Dutch literature see Stoop and Berg 1994.) Often, developments started in the U.S. were taken up later by these countries.

2. Like the notion of "scientific medical practice," "the medical profession" is of course no unitary whole. For the purposes of this chapter, however, I feel that this abstraction is warranted (cf. Abbott 1988).

3. As Harry Marks pointed out in a personal communication, "the intellectual and social programs of the 1970s might be seen to represent a fulfilment of the (failed) intellectual program of the Progressive era reformers in American medicine."

4. For the notion of coexistence see Mol and Berg 1994. The discourses I describe should be seen as ideal types: they are distilled from many different texts. Some articles contain subtle mixes; in others the typical elements of a discourse are only partially present. Further subdivisions can be made within some discourses, as will be seen in chapter 2.

5. On the effects of the success of medical research in World War II see Rothman 1991 and Marks 1988.

6. The distinction between "art" and "science" has always had its rhetorical use, in which the distribution of responsibilities and capacities is most prominent. To describe medical practice as an art emphasizes the importance of experience, of skillful mastery. This effectively ensures that the teaching of medicine remains under the direct control of the members of the medical profession. Likewise, the depiction of medicine as a science had its rhetorical function—see Abbott 1988 and note 10 to the present chapter. Anderson's (1992) fascinating story of the attempts to introduce computer diagnosis into the wards of an Australian hospital is a case in point. It would be erroneous, however, to reduce the usage of the terms "art" and "science" to this rhetorical function alone. That would bypass the great changes in the *meanings* of these terms over time. The changing meaning of "sci-

ence" with regard to medical practice is the topic here; however, it would be interesting to follow up on the close resemblance between the notion of "the art of medical practice" in the 1950s and the principle of specificity, which Warner (1986) discusses as having been typical of medical practice until the nineteenth century.

7. Also, the lure of financial gain sometimes may corrupt the physician. "Fee-for-service plans places a premium on keeping the patient sick rather than well," Oughterson (1955) argues.

8. This editor refers to the British National Health Service, which was a frightening example for the American medical profession. Lister (1957) wrote: "By being socialized, an art may become a trade . . . ; by socialization, a calling may become a business."

9. The misunderstanding, it is said, is often due to the many lay reports on medicine's problems showing "a regrettable failure to see both sides of the issue" (Anon. 1948c). Such was the case with the fire chief mentioned above, of whom the press had said that he had called 24 doctors and found none willing to come. After "a subsequent investigation," he appeared to have "actually talked only to two osteopaths" (Anon. 1948d).

10. The promise of science in postwar medicine should not be understood as resulting primarily from technically derived advances in medical care. The importance of the image of "science" for the strengthening of the professional identity and autonomy cannot be underestimated. In fact, this scientific promise was already shaping the medical profession's identity when it was as yet very unclear what a "science of medicine" actually was or could be—let alone that it had resulted in major clinical applications. For this latter point see Rosenberg 1987, Vogel and Rosenberg 1979, and Pasveer 1992; see also Warner 1985. The books by Rosenberg and Vogel deal with nineteenth-century and early-twentieth-century American medicine; Pasveer focuses on Britain and the Netherlands.

11. See Stevens 1989 and note 21 to this chapter.

12. For an elaboration of these usages see Willems 1995.

13. For a more recent, similar call, see Jencks 1992. The computer was both a reason for standardization and a means to standardize—consider the attempts, starting in the early 1960s, to devise an automated medical record (Anon. 1963a). On the fact that this electronic record has proved to be more perplexing than expected see Korpman and Lincoln 1988. See Kaplan 1995 for a brief history of medical computing.

14. The perception that "the mass of medical writing is expanding at an alarming rate" led to calls for standardization as well. A more standardized "classification of knowledge," Rutstein argued (1961), is a prerequisite for the physician to "maintain contact with medical knowledge." On the "information crisis" as a phenomenon of American science in the 1950s and 1960s see Wouters 1992.

15. This was not entirely new: in their attempts to "sell" randomized clinical trials to the medical professions, proponents of *these* rationalizing technologies often argued that physicians should allow such controlled experiments since physicians themselves, with each new treatment, were always already experimenting (Marks 1997; see also note 3 above). Analogies such as these had been made much earlier; for example, Flexner (1910) states that "the progress of science and the scientific or intelligent practice of medicine employ . . . exactly the same technique." See also Bishop 1900.

16. The problem-oriented record developed by Weed was much discussed and experimented with through the 1970s. In the 1980s these discussions quieted down somewhat, at least partly because of the developments described later in the present chapter. See Aring 1970 and Neellon 1974.

17. To attempt to achieve this, Weed later developed the Problem Knowledge Computer. On this project (and its small impact) see Weaver 1991.

18. Feinstein (1973c) used this example in a critique of Weed's plans. Dyspnea, properly used, indicates a shortness of breath of pulmonary origin.

19. See also Bleich 1971, LoGerfo 1977, and Feinstein 1987b.

20. In this discourse, the "art" of medicine is transformed too. When medical practice itself is addressed as a scientific activity, the art of medicine is transformed into a supplementary skill of adequately handling the personal aspects of care. This changing position is reflected in the fact that it was being addressed in a different, less solemn way. "The 'art of medicine' has been for long—perhaps too long—a rich source of 'corn' for the gristmills of medical orations. Yet it never has been clearly defined," Vaisrub argued in 1971. "Perhaps," he continues, "it should be sought in the poetry of physicians." This relocation effectively puts the art in a marginal position as an activity that only reflects upon medical practice.

21. See, for example, the call for "exactitude, clarity and precision" in Anon. 1922 and the applause for the development of standardized, pre-packaged drugs as a scientific achievement in Anon. 1926. See Stevens 1989 on the history of American hospitals in the twentieth century and Reverby 1981 on early-twentieth-century attempts to standardize record keeping. On the efforts to standardize diagnostic classifications (which go back to at least the nineteenth century) see Bowker and Star 1994.

22. While medical practice was being rendered a scientific activity, the general notion of "science" was steadily losing the indisputable status it had in the postwar years. Moser (1994) stated: "We *are* in a state of chaos, but it is simplistic nonsense to blame the mess on the clean, flourishing, nonmoral, logical and technical disciplines of science." Other authors were not so sure anymore. What had medical research meant for those concerned? Some research appeared to be only motivated by "the zeal to produce for the sake of recognition . . . and the urge to present a paper at a scientific meeting in order to qualify for an expense-free holiday" (Anon. 1960). Medical science may be non-moral, but it does have the capacity to visit "incomparable mischief upon mankind" (Ruby and Morganroth 1970).

23. See also Ingelfinger 1973. The usages of the term "autonomy" has just as many aspects as the usages of the terms "art" and "science" (see note 6 above). It is as much a rhetorical device for defending acquired professional privileges as a true concern of physicians regarding their service to their patients. For a detailed and powerful account of the medical profession's struggle for power and autonomy in the late nineteenth century and in the twentieth century see Starr 1982; see also Abbott 1989.

24. On the cognitive revolution in psychology see Baars 1986 and Boon 1982.

25. On the origin of the title see Elstein et al. 1990.

26. The problem of the increase of scientific data also affected medical education. In the 1960s and the 1970s, educational reformers pleaded for an educational approach which was less oriented to the amassing of facts. Medical students should learn "proper discipline in approaching medical problems." What was needed was a new educational philosophy, emphasizing the "primacy of process over content" (Elstein et al. 1978). Weed was one of the main spokesmen for these reformers. See Lave 1988 on the fundamental link between cognitive psychology and a positivist epistemology.

27. See also Wulff 1981, Smith 1988, and American Board of Internal Medicine 1979.

28. This is itself a simplification: different versions can be discerned within this variant as well, some focusing mainly on the physician as a statistical diagnostician, some focusing mainly on the maximization of utility. When looked at in more detail, what now appears as "one discourse" opens up in several, different approaches. Conversely, if I had strictly remained at the level of medical journal's editorials I would probably not have differentiated between the two discourses described in this subsection (see also chapter 2). All the quotes used here were explicitly descriptive.

29. See also Lusted 1971, Lusted 1975, Silverstein 1988, and Eddy 1990a.

30. Still other authors mix the two discourses, saying (e.g.) that the physician may fit each description, depending on the nature of the problem at hand. (See e.g. Szolovits and Pauker 1978.)

31. Abbott (1988) describes the shift in cultural legitimation for the profession's jurisdictional claims from "character" to more technically accounted means. Of course, instances can be found where authors talked about "scientific reasoning," or even "hypothetico-deductive reasoning" long before the cognitivist discourse came into being. Notwithstanding these isolated instances, however, it is only within this discourse that the notion of medical scientific action as a mental process becomes paramount.

32. See also Elstein et al. 1978 and Knaus 1986.

33. Physicians perform poorly in maximizing utilities, too. Elstein et al. (1986) found that even when physicians' estimated probabilities are reasonably correct,

they still did not reason consistently with them. Unaided clinical judgment, Elstein et al. argue, follows the principle of minimizing the most important risk, regardless of its probability. Politser (1981) summarizes findings from both the statistical and the information-processing approach. The question just how far these approaches can be merged in this kind of research is much debated in cognitive psychology. Kahneman et al. (1982), among the central authors in this debate, argue that they can and should be merged. With their work, they state, they can demonstrate that human judgments deviate from statistical theories such as Bayes' Theorem and can explain these deviations in the terminology of information processing. Gigerenzer and Murray (1987) vehemently deny that Kahneman et al. succeed in doing so. In the view of Gigerenzer and Murray, the information-processing terms Kahneman et al. introduce (e.g. the "heuristics" that explain certain deviations) are in fact nothing but statistical categories.

34. See also Lusted 1975 and Kassirer and Pauker 1978. On variations in medical practice see e.g. Wennberg 1984.

35. Compare McDonald 1976 and Kanouse and Jacoby 1988. It is not that Feinstein is not critical of the physician. Especially in his *Clinical Judgment* (1967), he is—and very much so. "If the clinician seems knowledgeable and authoritative, and if his reputation and results seem good, he can be condoned the most flagrant imprecisions, vagueness, and inconsistency in his conduct of therapy. The clinician does not even use a scientific name for his method of designing, executing and appraising [treatments]. He calls it *clinical judgment*." Feinstein is not so much blaming the *individual* physician, however. He does not localize the cause of medical practice's problems in the physician's mind (although he wrote a series of articles addressing the reasoning processes of physicians: Feinstein 1973a, 1973b, 1974). Rather, Feinstein blames the medical profession for not having generated a standardized and validated language. Under these conditions, there is not much an individual physician is to blame for.

36. Maxmen voices these ideas in rather strong terms. Other defenders of computer-based systems often stress that such systems cannot and should not replace the physician. This does not affect the argument here, however. Compare Barnett 1968, de Dombal et al. 1972, and Southgate 1975; see also chapter 2 of the present book.

Chapter 2

1. The empirical material used for this chapter consists mainly of journal articles dealing with the tools studied during the period 1945–1990, gathered by means of the "snowball method." In contrast with chapter 1, the editorials do not have a central position, nor do *NEJM* and *JAMA*. In addition, some of the key figures in the stories recounted here were interviewed.

2. Discourses cannot be disentangled from the (material) practices in which they have taken shape; vice versa, rationalities and views of medical practice can be inscribed in concrete tools. See Edwards's (1996) notion of "discourse"; see also Akrich 1992 and Hayles 1994.

3. See e.g. Anon. 1956b and MacMahon 1955. In his analysis of the history of the clinical trial, Marks qualifies the notion that statistics became so dominant a science for therapeutic research *because* it was seen as the route to medical truths. Rather, Marks (1988) argues, "to contemporaries, the improvements in experimental method offered by statisticians represented an elegant technical fix for a host of previously insoluble organizational and social problems"—as the problem to discipline individual physicians, the variations between methods and goals of involved groups, and so forth. (See also Marks 1997.)

4. See e.g. Warner et al. 1964, Gustafson et al. 1972, and Van Way et al. 1982. Contrary to de Dombal's tool, most of these other tools have never functioned in actual medical practice. Some used different statistical techniques—discriminant analysis, matching procedures, factor analysis, and so on. Bayes' Theorem, however, was used most often (see note 5). These specific differences are not vital to the arguments made in this chapter; those notions prevailed which came to be linked up with successfully developed tools. For overviews see Lusted 1968, Ledley 1965, Jacquez 1964, and Jacquez 1972.

5. Bayes' Theorem, it is argued, smoothly fits medical practice since it answers just the question a physician is looking for: what is the chance that this person, presenting with this and this symptom, has disease D (Ledley and Lusted 1959)? When we look at the probabilities figuring in the formula, these authors note that $P(S|D)$ is just the relation between symptoms and disease given by medical knowledge: this relation can be found in medical textbooks. It "depends primarily on the physiological-pathological aspects of the disease complex itself." $P(D)$, subsequently, is a factor incorporating local differences between populations. "This factor explains why a physician might tell a patient over the telephone [that a headache and fever probably indicate the flu, since it is around the community]. And the physician is more than likely right; he is using the $[P(D)]$ factor in making the diagnosis." (ibid.)

6. For this history see e.g. Edwards 1996, chapter 12 of Heims 1980, and Mirowski 1992. Mirowski convincingly shows how the link with the military shaped the form of current game theory. Porter (1995) traces the vocabulary of costs and benefits back to the U.S. Army Corps of Engineers' analyses of water projects at the beginning of the twentieth century.

7. Early pioneers were Ledley and Lusted (1959). See also Lusted 1968, Lusted 1971, and Schwartz et al. 1973. A classic textbook in this field is Weinstein et al. 1980. The technique can be applied by the individual physician (computer aids are available), or delivered by a "clinical decision consultation service" (Plante et al. 1986; Lau et al. 1983; Kassirer et al. 1987).

8. Some decision analysts and diagnostic tool makers would argue that they do not intend to say anything descriptive about the nature of the medical decision process. Most, however, do—either implicitly or explicitly. See Bell et al. 1988 and Gigerenzer and Murray 1987 for fundamental discussions of the not-so-clear line between descriptive and prescriptive models.

9. In the case of de Dombal's system, making the diagnosis *is* for a large part deciding upon therapy: when appendicitis is the most probable diagnosis, the patient should be operated upon; in most other diagnoses, less urgency is required. For an attempt to create a decision-analytic extension of de Dombal's tool see Clarke 1989.

10. For a study of the "framing effect" see McNeil 1982. Some authors doubt whether "subjective probabilities" and "utilities" are completely independent quantities (Fischoff 1988). "Sensitivity analysis" is often evoked as the solution out of the problems mentioned here. This is a method by which the input probabilities and utilities can be varied over broad ranges, so that one can see for which values the conclusions of the decision analysis hold. This added complexity, however, is only easily interpretable when the outcomes are clear-cut, i.e., when the conclusions of the analysis stay the same for a wide range of utilities and probabilities. More often than not, according to some, this is exactly the problem (cf. Doubilet and McNeil 1985).

11. In an offhand demonstration somebody once gave me, the best policy came out to be "not having the disease at all"— a sympathetic but alas hard-to-follow bit of advice.

12. In the eyes of some, this saturation with content explains the relative validity of many decision analyses. Shortliffe et al. (1979) argue that the usage of probabilities in a decision tree is less problematic than in a Bayesian diagnostic tool: the "powerful knowledge structure" of the former compensates for the imprecise subjective probabilities. In a tool like de Dombal's, however, "the 'knowledge' . . . lies in the conditional probabilities alone" (ibid.).

13. Of course, many "in-between" tools exist. Subjective probabilities, for example, have been used for diagnostic tools; see Gustafson et al. 1972.

14. See e.g. Eddy 1990c, Field and Lohr 1990, Komaroff 1982, and Komaroff et al. 1974. Emphasizing that protocols draw upon conditional (if . . . then . . .) rules, Lundsgaarde (1987) speaks of "noncomputerized expert systems."

15. It may be possible to trace links between Taylor's scientific management (see chapter 1) and the emergence of protocols in the work of Feinstein and Weed. Taylor's focus on the use of flow charts to represent work processes reappeared in the 1960s and the 1970s in the attempts of computer scientists to represent the flow of information in organizations (Friedman and Comford 1989, chapter 10). Feinstein (1974) comments that it is from computer science that he picked up the term "algorithm."

16. The first part of this title is derived from the title of Feinstein 1977.

17. Quoted in Vandenbroucke 1988.

18. Feinstein is by far the most outspoken and eloquent of these critics. Similar criticisms can be found in Cummins 1990, Margolis 1983, Ingelfinger 1975, and Rizzo 1993.

19. For example: the probabilities of being obese and having hypertension are not independent, since obesity increases the likelihood of high blood pressure. The independence assumption is required only when a simplified version of Bayes' Theorem is used, as is most often the case (Cornfield 1972).

20. Horrocks et al. (1972), introducing the Leeds abdominal pain system, themselves state that "unfortunately" their computer "is merely indicating that *if* the patient has one of the six or seven listed diseases in a 'database,' then the probabilities are as stated." In other words, the physician (or other health worker) first has to make a judgment whether the patient belongs to the population of "potential appendicitis patients." Only *within* this population do the stored probabilities have any meaning. Moreover, the computer cannot diagnose a patient having a disease which is *not* one of the seven diseases listed—it will, then, erroneously (and in the same vein as in regular cases) mention one of its seven options as the most likely diagnosis.

21. Feinstein is somewhat more sympathetic to the basic idea of decision analysis, since it incorporates some of the pragmatism and clinical sensitivity he fights for.

22. For a later, more extensive version of this protocol, see Winickoff et al. 1977.

23. Rheumatic fever is a potentially serious complication of streptococcal ear, nose, and/or throat infections, sometimes resulting in progressive formation of scar tissue on (and concurrent deformation of) the cardiac valves.

24. For some other contemporary examples of protocols for "physician extenders" see Sox et al. 1973, Komaroff et al. 1974, Grimm 1975, and Strasser et al. 1979.

25. Protocol builders often see no conflict between the design of protocols and the uniqueness of individual patients: through their branching logic, and through the leeway its rules allow, protocols can ensure or allow for individuation of the steps to be taken.

26. See also Field and Lohr 1990 and Eagle 1991.

27. See Mol and Berg 1994 for an analysis of "clinical," "statistical," and "pathophysiological" logics bearing family resemblance to the logics discussed here. See also the sociological analysis of the "clinical frame" by Dodier (forthcoming).

28. The criticism had a powerful appeal. The influence of the statistical ideal of objective inference was and is expanding. In the booming field of "quality assurance" it was also increasingly felt that deciding what proper medical action entails could not be left to local physicians. Objective, statistically processed "outcome figures" were needed to decide these issues (cf. Williamson 1973). See Gigerenzer et al. 1989 for acute observations of how statistics has come to "rule the world."

29. For an overview and evaluation of these American consensus conferences see Kanouse et al. 1989. More recently, the American Congress created the Agency for Health Care Policy and Research (AHCPR), with the development of "practice guidelines" as one of its functions (Field and Lohr 1990; Clinton 1992). On

the European experience see Vang 1988. In the Netherlands, "standards" for general practitioners are created by the NHG (Dutch General Practitioners Association), and guidelines directed more at specialist medicine are created by the CBO (Central Advisory Organ for Medical Audit). See e.g. Grol 1989, Zwaard et al. 1989, and Everdingen 1988.

30. Eddy (1990a) proposes such a combination of decision-analytic techniques and protocols into what he calls "practice policies." Clinical problems should be analyzed "in advance" using decision-analytic techniques, and the preferred steps to take should be written down in a practice policy.

31. For similar criticism from within the medical profession see May 1985, Rennie 1981, and Oliver 1985. Feinstein too (interview, November 11, 1993) has dismissed the "consensus" approach as "nonscientific."

32. Quoted in Arkes and Hammond 1986.

33. See Haugeland 1985. For an analysis of the links between the U.S. political discourse during World War II and the Cold War, and the origins of Artificial Intelligence (and the coming of the computer in general) see Edwards 1996. On the "metaphoric exchange" constructing the human in terms of the mechanical and vice versa see also Hayles 1994.

34. It should not be too explicated and clear-cut; that would make the task of designing the system trivial (Davis et al. 1977).

35. INTERNIST, for example, was an attempt to create a system which could make diagnoses within the broad field of general internal medicine (Schaffner 1985), and PIP (Present Illness Program) was to be a system which searched for the diagnosis in patients with edema through taking the history of the present illness (Pauker et al. 1976). For a detailed overview see Lipscombe 1991.

36. Szolovits and Pauker (1978) explain: "To the extent that anatomical and physiological mechanisms tie together many of the observations which we can make of the patient's condition and to the extent that our probabilistic models are incapable of capturing those ties, simplifications in the computational model will lead to errors of diagnosis." These authors also agree with Feinstein that in medical practice, the "basic premises of the applicability of Bayes' rule . . . are often violated." Since many patients have more than one disease at a time, the requirement that the hypotheses are "mutually exclusive" is often not satisfied, and errors in diagnosis will abound (ibid.).

37. Compare the critiques of Simon (1986) and Elstein et al. (1978).

38. It was, after all, their explicit focus on the substantive core of a domain which separated the expert-system builders from the earlier attempts to create artificial intelligence.

39. For some expert-system builders, the potential flexibility of expert systems also surpasses the possibilities of statistical tools (Pauker et al. 1976; Szolovits et al. 1988).

40. Compare Simon's emphasis on rationality as "adaptive behavior"—as behavior which is demanded by the environment (Newell and Simon 1972, chapter 3; Simon 1981). In his turn, Feinstein distrusts the expert system approach, criticizing it as an "academic" attempt to "bring outsiders' models in which do not work" (interview, November 11, 1993).

41. For an early, similar critique, see Kulikowski 1977. For a harsh critique on the "unscientific" nature of the whole expert system approach see White 1988.

42. Similarly, MYCIN used "certainty factors" to "permit a conclusion to be drawn with varying degrees of belief." In the rule described in the text, for example, the conclusion talks about "suggestive evidence," which stood for a certainty factor of (e.g.) 0.7. In combination with other rules, this allowed for "the accumulation of evidence" (Buchanan and Shortliffe 1984). Buchanan and Shortliffe argue why they found a more "formal" probabilistic theory like Bayes' Theorem both too limited and unfeasible, while also critiquing their own "groping effort to cope with the limitations of probability theory" (ibid.). Decision analysts also criticize expert-system builders for not paying explicit attention to the patient's perspective. Clinical decision analysis, according to Pauker (interview, October 29, 1993), is the only decision technique that offers a "hook" for the patient's preferences: "I don't know any other technique which allows shared decision making, that provides a formal process for combining the utilities, the values, the preferences of one individual with a formal structure and probabilities provided by another individual."

43. See e.g. Langlotz 1989, Shortliffe 1991, and Schwartz et al. 1987. See Gorry et al. 1973 on an attempt to create a computer-based decision tool with a decision tree built in. In such a system, however, as with the attempt to base protocols on decision-analytic outcomes, the "underlying decision models generally have been pre-specified" and "thus the program's usefulness is limited to those cases that correspond closely to the decision tree provided" (Shortliffe 1987).

44. For an attempt see Patil et al. 1982. On the general trend to elucidate the intricacies of the "clinical logic" further, see also de Vries and de Vries-Robbé 1985 and Reggia and Tuhrim 1985. See chapter 4 for additional reactions to some of the perceived problems of the first-generation expert systems.

45. More dimensions could be drawn out: the different tools contain different images of "the patient," they contain different images of the position of the physician relative to other health personnel, and so forth.

46. See also note 28. The term "Queen of Rationality" was coined by J. P. Vandenbroucke (interview, August 4, 1992). On the history of the clinical trial see Marks 1997.

Chapter 3

1. Following the work of Latour (1987), Callon (1991), and others, I conceptualize medical practices as networks of heterogeneous elements (including physicians, patient files, and x-ray machines)—elements which are interconnected in

manifold ways and which, taken as a whole, constitute the workplaces on which I focus. The heterogeneous nature of these processes is a theme that will recur several times in the remainder of this book. I speak of "negotiations" with nurses and laboratory tests, and I mention physicians, electrocardiogram machines, and auscultatory sounds as "spokespersons" for the heart. This terminology, probably odd to readers unfamiliar with recent studies of science and technology, is an attempt to keep a more symmetrical perspective. One of the central tenets of these studies is that the development of a technology cannot be properly understood from perspectives that treat nature and society as separate realms and confine explanatory power to either one of them. Exactly what the transformation of a practice (or of a tool) comes down to can never be understood from standpoints that center on the nature of a medical problem or, conversely, on the actions of medical personnel. Rather, the physical and social aspects of "computerized" or "protocolized" medical practices are the *outcomes* of the historically contingent events that have led to their current configuration. Only by treating them symmetrically can we hope to gain an understanding of the "success" or "failure" of decision-support techniques; only by focusing on the way these heterogeneous networks take shape and break down can we come to terms with the fundamental issues at stake in the production and use of technical systems in medical practices.

2. The reason I have not focused on clinical decision analysis is that this technique is very rarely used in the form as described in chapter 2 (see also chapter 6). The stories I tell about the tools are based on articles their builders have written about them, on observations of the tools, on observations of the processes of tool construction (in the case of the research protocol), and on interviews and correspondence with the tool builders.

3. This therapy is a recent alternative to the transplantation of bone-marrow cells harvested from the patient's pelvic bone. This new technique, it is claimed, is more convenient for the patient and results in faster restoration of normal blood-cell levels.

4. As both Wyatt and Emerson remarked, Emerson had been thinking of building a system like ACORN for some time already. The first few paragraphs of ACORN's story, as depicted here, constitute an origin myth that could easily be told otherwise. (One could argue, as Wyatt did, that the program started out as an effort to reduce the number of tests done in the emergency department.) The problem, however, is that *each* alternative could be told otherwise—such is the nature of origin myths. Since each of the issues mentioned did, at some stage, play an important role, and since the arguments of this chapter are not dependent on the specific origin myth chosen, I chose to keep the first paragraphs of the story more or less as Wyatt (1989) recounts it.

5. Here, "could" implies not just the ability to gather the item, but also the legal and professional issues involved. Nurses, for instance, could not be given the responsibility to judge an electrocardiography report—something which, in practice, experienced nurses often do much better than junior physicians (see e.g. Hughes 1988). On the skill of reading electrocardiograms see Hartland 1993b.

6. "Repeatability" was measured by the mean percentage agreement between a first and a second elicitation of the data item; in the case of a yes/no question, of course, 50 percent repeatability is expected by chance.

7. A test set of 133 patients had been kept apart from the set with which the Bayesian formula had been devised.

8. The rules were predominantly directed at "further reasoning about the ECG and about specific combinations of data," for example about data that might indicate unstable angina. See Wyatt 1991a for a more detailed description of ACORN's rule base.

9. Electrocardiography yields diagnostic information about the rhythm of the heart, and about whether (e.g.) the heart seems to be ischemic.

10. A technoscientific script is not simply equal to the prescriptions in the protocol. The interests at stake, the redistribution of costs, the research careers involved, the technologies which are skipped, the laboratory tests which are deemed more crucial than others: these issues are not explicated in the text, yet are of prime importance for an understanding of the development and use of the tool. See also Timmermans and Berg 1996.

11. See Latour 1987 for the notion of "spokesperson"; see also note 1 above.

12. The names in this case fragment are fictional.

13. Hence the term "heterogeneous engineering," coined by Law (1987).

14. Not all formal systems contain a pre-fixed list of statements as their outcome. Decision analysis does not: there, the outcome is a set of numbers, generated through the application of the decision-analytic method on the input data (see e.g. Weinstein et al. 1980). The outcome is just as much pre-defined within the rules, however: it is always a similar set of numbers.

15. For a study of the constitution of these new immunological entities see Cambrosio and Keating 1992.

16. On similar issues in MYCIN see e.g. Lipscombe 1991. Davis (1977), in discussing MYCIN's rule base, states that a domain should have "a limited sort of interaction between conceptual primitives." If more than six clauses start appearing in the premises of a rule, the "premise becomes conceptually unwieldy." In other words, when too many factors can impinge upon a relation between a data item and, say, a given outcome, these relations become practically infeasible to hold on to. Similarly, he argues, it is needed that "the presence or absence of each of those factors can be established without adverse effect on the others"—otherwise the results will "depend on the order in which the evidence is collected."

17. Szolovits and Long 1982. See also Schwartz et al. 1987; Engle 1992; de Dombal 1990.

18. The term "digitization" is from Haugeland (1985). Collins (1990) also takes up this notion.

19. See also Horrocks et al. 1976 and Wilson et al. 1977. De Dombal's tool could not work with data items that took time to gather. Like ACORN, this tool dealt with emergency situations: data items could only be used if they could be collected and entered quickly. Unlike Emerson's team, however, de Dombal and his coworkers had physicians doing the investigation. In this way, more physical examination items could be used. Also, it appeared that with these data items enough stable relations could be drawn to cover the majority of acute abdominal pain patients coming to the emergency ward. Here again, it was only by training the physicians, introducing specifically developed forms, and hiring research assistants that they were able to do so. Here again, the tool could work only once a very diverse and entangled range of elements were disciplined.

20. Parallels to the analysis given here can be found in Marks's work (1988, 1997) on the coming of the clinical trial. The analysis draws strongly on the insights gained in the studies of the stabilization of scientific facts and technologies: see e.g. Latour 1988, Shapin and Schaffer 1985, and Bijker and Law 1992. See chapter 1 of Zuboff 1988 for a historical background to the depth and scope of disciplining practices. (Zuboff focuses primarily on the disciplining of workers.) See Woolgar 1991 for a study of how users are configured. Coming from an entirely different (cybernetic) background, Beniger's (1986) analysis of heterogeneous technologies of control yields conclusions interestingly similar to those drawn here.

21. In the case of decision-analytic techniques, of course, patients' preferences are central—but here too the patient's role is highly pre-structured. (See chapters 2 and 4.)

22. See e.g. Fujimura 1992 and Latour and Woolgar 1986. See also Horstman (forthcoming) on the construction of reliability in the case of the urine test in insurance medicine. A beautiful attempt to come to terms with the way matter is *not* regularly tamed is Haraway's (1991) discussion of nature as "coyote/trickster."

23. Among these institutions are pharmaceutical industries and local, national, and international research associations. In fact, these links to interinstitutional and international protocols partially explain why (for example) oncology wards look so alike all over the Western world—much more alike, often, than different wards within one and the same hospital (Fujimura 1987, 1992; chapter 6 of present volume). See also Timmermans and Berg 1996.

24. See Bowker's (1994b) notion of infrastructure. See also O'Connell's (1993) study of the creation of universality through the circulation of particulars.

25. This simplicity, it should be clear, is a *consequence* of the work performed to achieve this robustness—what O'Connell (1993), after Latour, calls "metrological practices."

26. Note that there is no definite way to ground such a judgment. There is no way to determine what is the "more truly indicative" set of data items, for example. There is no solid, independent ground from which to make some ultimate judg-

ment on a new selection of indicants. Hence, redelegating spokesmanship will result in subtle but thorough transformations in disease definitions and criteria. See Lynch 1985, Latour 1988, Pasveer 1992, and Hirschauer 1991.

27. Feinstein's life work could be summarized in this sentence—see e.g. his *Clinimetrics* (1987a).

28. See note 1 above.

Chapter 4

1. This included both reactions due to the high osmolality of the iodinated contrast agents and allergic reactions. The non-ionic agents supposedly caused fewer adverse effects in both categories.

2. The ionic agents were cheaper than the non-ionic ones.

3. This is the "framing effect" mentioned in note 10 to chapter 2.

4. Interview, Peter Emerson July 21, 1993. The work of Bowker and Star (1994) on the International Classification of Diseases has been especially influential in the writing of this section; see also Fujimura 1987, Fujimura 1992, Star 1989, Star 1991a, Star and Griesemer 1992, Timmermans et al. (forthcoming), Kling and Scacchi 1982, and Webster 1991. Studies of information technologies in medical domains in which the same phenomena as described here can be seen to occur include Dent 1990, Rudinow-Saetnan 1991, Bloomfield 1991, Lipscombe 1991, and Hartland 1993a. The latter two cover a whole range of different medical expert systems.

5. This may sound somewhat cryptic, but the statistical method by which the tool builders selected the "powerful" from the "weak" data was based on a database containing only a sample of the *local* population of "patients with chest pain coming to this emergency ward."

6. When asked about the definition of acute abdominal pain, de Dombal replied in 1976: "Of course everybody else has their own definition, and this may differ from place to place. At the moment, we don't even know what these differences are—but we must certainly find out." (de Dombal and Gremy 1976) In building computer-based decision tools, many authors have come up against this issue as a hindrance to their goals (see e.g. Barnett 1968 and Lusted 1968). The following quote from de Dombal 1989 illustrates the persistence of this problem: as the first "obstacle to progress, [standing] in the way of widespread implementation," de Dombal mentions the lack of clear medical terminology. "Medical questions and answers have to be simplified, pre-defined, and agreed in advance by wide consensus," he argues (see also Gill et al. 1973).

7. See Bjerregaard et al. 1976, Horrocks et al. 1976, and Fenyo et al. 1987. "The clinical spectrum of e.g. pancreatitis cases differs," Bjerregaard et al. note, "alcoholic male cases being seen more often in Copenhagen than in Leeds."

8. Similar problems have been reported for consensus reports. Kosecoff et al. (1987) found that compliance with consensus statements varied by hospital. One of these statements, regarding urgent performance of coronary angiography in patients with unstable angina, was (not surprisingly) found to be adhered to more in hospitals having catheterization laboratories. See also Lomas 1989.

9. Several authors have reported that physicians often see consensus reports as "biased." According to Hill and Weisman (1991), "the physicians seem to be aware of the sponsors' intent to produce a document that meets the sponsors' need to communicate a particular message about how medicine ought to be practiced." See also Grol 1989.

10. Marks's (1997) account of the therapeutic research on pneumonia serum and arsphenamine at the beginning of the twentieth century provides a historical example of this phenomenon.

11. Interview, de Dombal. See also Ikonen et al. 1983, de Dombal 1988, Chatbanchai et al. 1989, and de Dombal et al. 1991.

12. Field and Lohr 1990; Eddy 1990e; Eddy 1990f; Society for Medical Decision Making Committee on Standardization of Clinical Algorithms 1992. Eddy's difficulties to use international data were due to an array of reasons resembling the "non-transferability of databases" discussed above.

13. Intimately related to localization in space is what might be separated as yet an additional type of localization: *localization in time*. A tool is always bound to a specific time, to a specific state of medical knowledge, a specific epidemiology: just as it is located in a space, it has a history and a projected future. (See the notion of the trajectory of decision-support tools in Timmermans and Berg 1996.) I have not opted for localization in time as a separate category however, since striving for a tool which has an infinite, universal usability through *time* is generally not part of the advocates' ideal-typed views.

14. For similar self-diagnoses of the "current state of the art" in computer-aided decision tools see Miller and Masarie 1990 and Schwartz et al. 1987. A typical example is MYCIN's successor, ONCOCIN, which explicitly "retreated" on a very small domain: it is designed to control the use of pre-determined, oncological protocols (Shortliffe and Clancey 1984). A more detailed discussion of ONCOCIN can be found in Lipscombe 1991; see also Berg 1994.

15. Wilson et al. 1977; interview de Dombal. For similar experiences see e.g. Davis et al. 1977 and Engle 1992. If nothing more, it is an enormous amount of work just to expand the required data base and/or rule set: INTERNIST, designed to cover "all of internal medicine," had cost 15 person-years of programming time. INTERNIST, however, was notorious for its strange interactions between rules, its many errors, its impracticality, and its overall sacrifice of depth, precision, and transparency (Miller et al. 1986; Miller and Masarie Jr. 1990; see also Sutherland 1986).

16. Similarly, Greenfield's protocol for upper-respiratory-tract infection cut off some diagnostic routes since the health assistants were not trained to examine the

heart: it was not contemplated to have them listen to murmurs or gallops (interview, Anthony Komaroff, November 9, 1993).

17. I am not arguing that no system could do so. Rather, the point is that *in this specific practice*, with this set of data items available, this local population of patients, this group of nurses, and so forth, it was not feasible.

18. "Scope" has a temporal dimension too: many tool builders have criticized their own tools for just taking "snapshot" pictures, focusing merely on a "'snapshot' of data about a patient at a fixed time" (Buchanan and Shortliffe 1984). Most computer-based tools do not take the temporal development of illness, symptoms, reactions to therapy, and so forth into account; "the past is treated in the same vein as the number of years you've smoked" (discussion with Bruce Buchanan, December 6, 1993).

19. On this phenomenon as a broader trend see e.g. Perolle 1991.

20. Some commentators have argued that just about any tradeoff (say between the speed of the tool in use and its comprehensiveness) could be added as an additional type of "localization." I do not think that this is so. I focus on space, scope, and rationale, since these are dimensions where *universality* is the ultimate ideal-typed goal *and* the projected reason of existence of these tools. "Comprehensiveness" or "speed" is much less basic to what these tools stand for than, e.g., "scope."

21. For a similar tale about policy-oriented decision analysis see Ashmore et al. 1989.

22. Again, this is a phenomenon that can be seen time after time. On MYCIN see Buchanan and Shortliffe 1984; on INTERNIST see Miller et al. 1982, Pople 1982, and Miller et al. 1986.

23. The expression "light without heat" is derived from Zuboff 1988. For a harsh insiders' critique of the success of expert systems see Sutherland 1986.

24. See also Löwy (forthcoming). For example, joining the breast cancer protocol group had a distinctive advantage for oncological centers: they would get additional funds with which they could buy shiny state-of-the-art technology. On the status attached to working with decision-support technologies see Ashmore et al. 1989, Bloomfield 1991, and Saetnan 1991.

Chapter 5

1. My field experience in medical practices started when I worked as an intern. In addition, I spent two months on an oncological ward specifically studying protocols in use. For the other tools see note 2 to chapter 3.

2. For a study investigating what is it to build a tool (in this case, an expert system) see Suchman and Trigg 1993.

3. On the links between cognitive psychology and a positivist epistemology see Lave 1988 and Gigerenzer and Murray 1987; see also chapter 2 above. See also the work of Forsythe (1992; 1993a; 1993b) focusing on the inscription of positivist interpretations of "work," "information," "knowledge," and "explanation" in medical expert systems. This critique has its analogies in many fields—see e.g. the influential early criticisms of policy-decision-making models of Braybrooke and Lindblom (1963) and Allison (1971), the critical appraisal of economic decision analysis by Ashmore et al. (1989), the analysis of automated administrative systems in concrete work settings (e.g. Gasser 1986), and criticism of automation in the military (see Edwards 1996).

4. See e.g. Lipscombe 1989, Hartland 1993a, Hartland 1993b, Dreyfus and Dreyfus 1986, and Gordon 1988.

5. Lave (1988) talks about "problem management" as opposed to "decision making" or "problem solving." Both of the latter have strong cognitivist connotations.

6. Bone-marrow transplantation can be performed with a donor's marrow, in which case it is called *allogenic*. In this case Mr. Wood's own marrow was used—an *autologous* transplantation.

7. For Strauss et al. (1985), the notion "illness trajectory" includes the course of the illness, the total organization of work done over that period, and the impact of that work on those involved. See also Silverman's (1987) concept of "site."

8. Anamnestic information is information obtained through questioning the patient, also called the patient's "history." Some studies showing the constructed nature of anamnestic data are Cicourel 1986, Davis 1986, Silverman 1987, Helman 1988, A. Wynne 1988, and Fisher and Todd 1983. For studies depicting medical practice as a locus of constructive *work* see Bloor 1976, Bloor 1978, Lynch 1984, Hirschauer 1991, Atkinson 1981, and Atkinson 1995. See also the studies reported by Casper and Berg (1995) and by Berg and Mol (forthcoming).

9. One could object that these physicians do not listen well or are biased toward a pre-supposed diagnosis. Such objections, however, still assume that historical data are there to be found by a physician who listens properly. As stated, however, historical data are not "givens" but are produced in and through the doctor-patient contact. In that setting, the historical data take shape for both the physician and the patient (see the studies quoted in the preceding note).

10. On the production of realities in medical work see Hirschauer (forthcoming) and Cussins (forthcoming).

11. On the illusion of the "true image" see Mol (forthcoming).

12. See Berg 1992 for a more detailed exposition of this argument. For a superb account on how accounts of objects are continually modified in laboratory "shop talk" see Lynch 1985.

13. See note 26 to chapter 3.

14. Garfinkel (1967) phrased it powerfully: "For the practical decider the 'actual occasion' as a phenomenon in its own right exercised overwhelming priority of relevance to which 'decision rules' of theories of decision making were without exception subordinated in order to assess their rational features rather than vice versa." For similar accounts of rules and their use see Zimmerman 1970, Heritage 1984, Knorr-Cetina 1981, Lynch et al. 1983, and Harper and Hughes 1993. See also Lynch 1993 on how Wittgenstein's (and ethnomethodology's) views on rules should affect science and technology studies. On the metaphor of "fluids" see Mol and Law 1994.

15. Compare Star 1989a and Forsythe 1993b. Clinical decision analysis, of course, does take the patient's preferences into account (see chapter 2).

16. See Timmermans 1993 for the notion of "crystallization points" in the ongoing flow of medical work.

17. The complexity of this work is deepened by the fact that many patients have more than one affliction—and thus more than one illness trajectory—and by the corresponding organizational work.

18. These wordings are after Lynch (1985) and Knorr-Cetina (1981).

19. At the same time, the avalanche of information that piles up in the underlying, forgotten pages can continually be reaccessed and reinterpreted. Their sheer volume and scope allows endless remodifications and rewritings of histories that once seemed to be closed. Anomalies are easily found, and the past is easily rewritten in the light of the unfolding present—and vice versa.

20. See Latour 1988 and Garfinkel 1967. On the temporal structure of medical work see Zerubavel 1979.

21. On "cognition" as a social process see Woolgar 1989, Coulter 1983, Coulter 1989, Amann and Knorr 1989, Young 1981a, Young 1981b, Latour 1986, and Hutchins 1995.

22. On the "in-course" accomplishment of a trajectory's itinerary see Luff and Heath 1993, Knorr-Cetina and Amann 1990, Knorr-Cetina 1995, and Suchman 1993a. For similar comments on the notion of "decision" see the studies of decision making in neonatal intensive care units by Frohock (1986) and Anspach (1993).

23. In actor-network terms: a part of the network is collapsed into one node (see e.g. Callon 1987).

24. In addition to making complexity doable and/or enhancing comparability, a tool may be intended simply to standardize sequences of activities or to process a set of data to some pre-set ideal. (See the protocol for chronic lung disease in figure 2.10.)

25. In addition, taking the tool's history of localization into account already problematizes the idea of a tool in "control."

26. See note 4 above.

27. This may be a designer's choice, but it need not be—it may just be the way a particular implementation process has evolved.

28. I focus on breakdowns since this is where the issues at stake are most prominently visible—see Garfinkel's (1967) methodological recommendations. Reinterpreting the moments portrayed as "anecdotal exceptions to the rule" misses the methodological point that the empirical case fragments are not what "proofs" the argument. They *illustrate* the argument, and *clarify* it—the "proof," however, is in the persuasiveness of the argument itself.

29. Misunderstanding (in both directions: the patient misinterpreting the computer and vice versa) occurred easily, and this was upsetting to patients who had no clue of what the consequences of such misunderstanding might be. Patients could also become upset when the computer would *not* ask a certain question. A woman became distressed after the interview was over, saying: "It doesn't ask you if you have had gallstones. I have had my gallbladder out, but it didn't ask me that, so this interview doesn't show what I am really like." (Hartland 1993a) For an analysis of similar troubles in human-machine communication see Suchman 1987.

30. Collins (1990) offers a hilarious but illustrative thought experiment in which an expert system designated to deal with complaints of acute pain (say, ACORN or de Dombal's tool) is confronted with a patient with a pointed weapon embedded in his body. The tools are not programmed to ask "Is there anything stuck in your back?"—they will just run their pre-set questions and generate one of their pre-set diagnoses. It is the task of medical personnel to "repair" such errors.

31. "Protocol violation" is an expression used by members of a medical staff. Whether an event is a violation, and, if so, whether it is "major" or "minor," is an ongoing matter of debate (Bosk 1979; Lynch 1985).

32. These staging definitions are generally included as an appendix in oncological research protocols.

33. In such situations, Suchman (1987) states, these rules "pose problems of interpretation that are solved in and through the objects and actions to which the instructions refer." The focus on "negotiated interpretations" is a core tenet of the "controversy" studies in the sociology of science, exemplified by Collins 1985 and Mulkay 1988. See also Woolgar 1988.

34. Likewise, personnel working with a system evaluating measurements in patients with peripheral vascular disease (Talmon et al. 1987) ran into troubles. Ordinarily, they would deem blood-pressure measurements unreliable whenever they were "too high"—something, they explained, that could be due to the blood vessels' being arteriosclerotic. Since the system had no specific provision for such situations, entering these measurements into the system would obviously set it off into a wrong direction. Not entering anything, however, would be interpreted as

entering "zero," which would be equally wrong. The staff members involved had no clue how to reinterpret the tool's demands in such a situation.

35. A similar phenomenon would have plagued MYCIN. Buchanan and Shortliffe (1984) state that "MYCIN's control structure is not concerned with resource allocation; it assumes that there is time to gather all available information that is relevant [it would ask 20–70 questions—MB] and time to process it."

36. See e.g. Boey and Dunphy 1985.

37. As an obvious but illustrative example, consider the fictitious patient mentioned in note 31 entering the emergency department. The tool's diagnostic output is limited to seven non-traumatic categories (see chapter 2), so filling in the form and translating "pain in the back" as "pain in the flank" will here ensure the generation of a meaningless diagnosis. On the effort to "align" oneself with a tool see Suchman 1987 and Wooffit and Fraser 1993. This is an instance of a system's "falling off the knowledge cliff" (Forsythe 1993a)—a problem of unanticipatedly limited scope.

38. On the distribution of knowledge in people performing different subtasks of a same overall task see Hutchins 1995. For a nice example of how nurses' knowledge of physician's subtasks is often crucial see Hughes 1988.

39. Owing to its black-box character, this issue is particularly salient for computer-based tools. A protocol is always completely explicated on a piece of paper. Any staff member with a knowledge of the protocol's domain can at least form some judgment as to the protocol's purposes and potential bottlenecks—as, in fact, is done continually.

40. See also Jordan and Lynch 1993, B. Wynne 1988, Barley 1988, and Callon 1980.

41. Whether in space, in scope, or in rationale (or any combination of these) is an empirical question.

Chapter 6

1. This number excludes four "educational systems" and some statistical tools (including de Dombal's tool). It includes, on the other hand, several tools which are not primarily designed for usage in medical practice (such as a system which reviews physicians' prescribing patterns for Medicaid patients' drug utilization). The list (version V1.5, July 21, 1993), compiled by Enrico Coiera, appears on the AI-Medicine mailing list (via e-mail: ai-medicine-REQUEST@med.stanford.edu). The following systems are in operation in more than one location: two laboratory systems for analysis of test results, PUFF (a system which interprets pulmonary function data in lung function laboratories (Aikins et al. 1983)), APACHE III (which predicts the risk of dying in the hospital for admitted patients (Knaus et al. 1991)), and a "managed second surgical opinion system."

2. On MYCIN see Buchanan and Shortliffe 1984 and Lipscombe 1991. On INTERNIST's "demise" and transformation into QMR see Miller et al. 1986 and Miller and Masarie 1990.

3. See also Balla et al. 1989. See also the poor results of a "clinical trial of clinical decision analysis" performed by Clancy et al. (1988) in an attempt to rationalize physicians' vaccination decisions (for themselves) through decision analysis.

4. On the limited practical success in the Netherlands see Knottnerus 1987 and Warndorff et al. 1988. In health policy, clinical decision analysis appears to be more successful—but see the critical appraisal of Ashmore et al. (1989).

5. On the explosion of research protocols in postwar medical practice see e.g. Löwy (forthcoming) and Patterson 1987. See Löwy (ibid.), Richards 1991, Epstein 1995, and Marks 1997 for studies of the negotiation processes typifying the construction and evaluation of clinical trials.

6. See also Lomas et al. 1989. Hill and Weisman (1991) draw similar conclusions in their prospective study of Maryland physicians: ". . . the strongest predictor of congruent practice behavior 1 year after [the report] is congruent practice behavior just prior to the report's release."

7. See also Sox et al. 1973.

8. This is not to say that these advocates claim that clinical decision analysis has to be applied always; many authors state that it should be reserved for "difficult cases" (Schwartz et al. 1973).

9. According to Stephen Pauker, one of the founders of clinical decision analysis, the main problem is that there is no niche in the financial organization of medical care: "It is an activity which is not reimbursable by insurance companies or whoever pays for medical care. . . . We are only reimbursed for seeing a patient. For contact time, or for procedures. I am a cardiologist. If I do a cardiac catheterization, and spend an hour and a half doing it, the insurance company would pay me $1500. If I do eight hours on the analysis of a patient, if they would pay me at all, they'd pay me for medical consultation, which could pay out at $100." (interview, October 29, 1993) See also Littenberg and Sox 1988, Detsky et al. 1987, and Dolan 1990.

10. In principle, advocates argue, wherever humans apply general knowledge to a particular problem, computer-based decision-support systems should be feasible. Nevertheless, the practical state of expert-system research might force a restriction to relatively well-defined domains—such as medicine (Duda and Shortliffe 1983).

11. See also de Dombal 1987b. For earlier analyses of the limited success of computer-based decision tools see Duda and Shortliffe 1983 and Barnett 1968.

12. See also Kahan et al. 1988.

13. See also Kanouse and Jacoby 1988, Lomas et al. 1991, and Zwaard et al. 1989.

14. See Böckenholt and Weber 1992 for such an analysis of the general category of "formal methods in medical decision making." This view is shared by "diffusion" sociologists, who attribute the failure of computer-based decision tools to catch on to "too little information to doctors," "negative attitudes of medical personnel," and so forth. See e.g. Anderson and Jay 1987, Lundsgaarde 1987, and Weaver 1991. Evans (1990) argues that to understand the success or failure of expert systems one should look at the medical profession's "successful suppression of the nurse practitioner": it is all a matter of "control over the social and economic context of practice." On this mode of explaining "failure" see Kling and Iacono 1984, Kaplan 1987, Kaplan 1995, Star 1989a, Star 1989b, and Forsythe 1992.

15. See e.g. Yu et al. 1979. On the evaluation of expert systems in medicine see Lundsgaarde 1987, Wyatt 1991a, and Miller 1986.

16. Alternatively, advocates sometimes argue that "intimacy" and abstraction are not on the same continuum after all; that formalizing the "technical" aspects of the work *creates room* for the "uniquely human" skills as "the management of the emotions" (Schwartz 1970).

17. Fuller (1993a) criticizes the "practice-mysticism" of authors, such as Collins (1985, 1990), who argue that there is some "tacit dimension" in most domains of human action (yet another foundation) that will forever elude all efforts to formalize it.

18. Woolgar (1985, 1987), Turkle (1984), Hayles (1994), P. Edwards (1996), and others have addressed how new technologies transform our notions of what it means to be human. See also the articles in Ashmore et al. 1994. Contrasting human practice and formal tool, comparing their "performance" on "similar tasks," already contributes to the active process of rendering them equivalent. Debating the superiority of human expert decision making over formal tools' decision making (or vice versa), for instance, implies a highly individualistic and mentalistic understanding of what "expert action" (in this case, medical work) is (Fuller 1993a,b). The question of superiority or equivalence is to be settled as if a game of chess, one lone individual posited versus another, is paradigmatic of what medical work comes down to (see also D. Edwards 1994).

19. On the digitalization of medical practices see Anon. 1964 and Reiser 1978; for an account of the highly digitalized practice of current neonatology see Mesman 1993. This tendency to quantify information is obviously tied to the current high status of the statistical logic discussed in chapter 2 above.

20. For example, MYCIN's successor, ONCOCIN, was explicitly intended as a tool to be implemented in a practice already "predisposed" to formal tools. ONCOCIN is supposed to help oncologists follow research protocols. See Shortliffe et al. 1984, Berg 1994, and Lipscombe 1991.

21. See Marks 1988 for an excellent illustration of this point in the case of research protocols.

22. Research on the treatment of cancer was one of the main spearheads of the coordinated medical research effort that boomed after World War II. See Löwy 1995, Marks 1988, and Zubrod 1984.

23. This rendering is not completely fair to at least one of those I have labeled critics. Collins can be said to look at the process of getting a tool to work. His position is a social constructivist's inversion of Dreyfus. He argues against Dreyfus that whether a domain is formalizable is not a given feature of the natural world out there. Even mathematics and physics, Collins argues, are formalizable only because in the social practices constituting these domains we have *opted* to behave in a formalizable way. Collins calls this "behavior specific action," and he sees it as a specific, willfully created subcategory of human "regular action." Regular action is itself fundamentally unformalizable. In turning Dreyfus upside down, however, Collins ends up with a position that is similarly troublesome. Now the formalizability of a domain is no longer a given property of the world out there; it has become a property entirely situated in the domain of social phenomena. Translated into the vocabulary deployed here, Collins points at the need to discipline a practice to a formalism—but he focuses solely on the need of *people* to conform themselves to the tools' demands. Overlooking the sheer heterogeneity of these practices, the assembly of bodies, artifacts, machines, people, and so forth, which *all* need to be disciplined, Collins in fact does not bring Dreyfus's analysis much further. In addition, since it makes the *a priori* distinction between "regular" and "behavior specific" action, Collins's argument is just as foundationalist as Dreyfus's position.

24. For the term "obligatory passage point" see Latour 1987.

25. My characterization of "formal tools" as requiring constraints on input, output, and the relation between data entered and output statements (chapter 3) and my distinction between "accumulating" and "coordinating" tools (chapter 5) both already contain this interrelation of "tool" and "context." These definitions are useful as far as the events described in this book go (i.e., they are not simply an alternative set of essential properties), and they describe the tool in terms of its relation to the network in which it is to be embedded.

26. These comments feed directly into the current "rewritings" of actor-network theory. A broad range of authors have pointed out that this framework tends to focus unduly on the exertions of the central actor, that it embodies an engineering logic (building wholes from heterogeneous elements, controlling the periphery from the center); that it, in Haraway's (1994) terms, produces yet "another Sacred Image of the Same" through its metaphors of "heroic trials of strength" (see also Star 1991 and Lee and Brown 1994). In other words, actor-network theory might embody too many echoes of the modernist constitution it tries to overcome (Callon and Law 1995; Law and Mol, forthcoming; see also Timmermans and Berg 1996).

27. For a eloquent critique of the notion that techniques are devoid of being see Latour 1993.

28. This is and can only be a conjectural claim: there is no way this could be empirically "validated." Even if two systems could be found that dealt with *exactly* the same clinical problem (and given their different views of what medical practice's problems are, this is rare), it always remains possible to argue that differences in advice found are due to differences in assumptions embedded in the rules and formulas which are *not* related to the different rationalities investigated.

29. The requirement of human intervention to overcome the ineptness of machines in the social order is the theme of Collins (see note 23 above).

30. Here I paraphrase Robinson (1994).

31. Robinson's (1991b) notion of double-level languages points at this interlocking of the formal and the informal.

32. The two need not fully overlap: clinicians might think there is a high chance of appendicitis but—for a variety of reasons—delay surgery anyway. On the other hand, they might estimate the chance of appendicitis as rather low, but operate anyway—when, for instance, they are hesitating between "appendicitis" as a probable diagnoses and another affliction that would also require a similar, urgent surgical intervention.

33. The processing of its raw data involves many "decisions" on the way: the tool draws upon a specific mathematical technique in reconstructing the composite image from the individual shots taken and not on others, it makes a certain number of readings, and it represents its calculations pictorially, drawing on specific algorithms (Blume 1992).

34. Of course, this issue is not limited to situations where decision-support tools are at stake—see e.g. de Vries 1993. The matter just comes up more saliently here.

35. This wording is after Lynch (1993).

36. The evolution of de Dombal's tool leads to a similar conclusion. This tool started out in the early 1970s as a diagnostic machine—as a tool that would offer the most probable diagnosis given a case of acute abdominal pain. De Dombal's team produced many papers showing how badly physicians performed on this task, and how much care improved when their tool was used. Over the years, however, de Dombal and his team started to realize that "at least half of the improvement [in clinical performance] was because of the adoption of preagreed, predefined medical terminology, coupled with discipline in data collection" (de Dombal 1987a). Some years later, a study demonstrated that the results witnessed were due entirely to the structured data collection ensured by the form—whether the physicians ever saw the computer's advice or not did not make any difference (de Dombal et al. 1991). Again, it seems that the cognitivist discourse, of which de Dombal's statistical tool was a paradigmatic exemplar, provided a wrong tool for the job: the paper form would have yielded the same results by itself. Again, a solution was provided for a problem that did not exist at the level of the work of medical personnel. The problem was not so much a suboptimal mental functioning of physicians as a matter of having too little time

for a patient and of not being comprehensive enough in one's physical examination and history taking.

37. This is not the place to dwell in more detail on how "design" would have to be refigured. For some inspiring and much more elaborated accounts of alternative design methodologies see Zuboff 1988, Greenbaum and Kyng 1991, Robinson 1991a, Robinson 1991b, and Agre 1994.

38. At the very least, the fact that so many rationalities coexist is a strong argument for participation of a broad range of potentially affected groups (such as patient organizations) in the construction of new decision-support techniques—including, importantly, the research protocol.

39. See also Zuboff 1988. Focusing on this aim would be a wholly different (and potentially more fruitful) approach to enhancing a profession's status.

40. Sometimes, discussions in the literature of computer-supported cooperative work seem to aim toward a desire for technologies that are like blank sheets of paper—that are totally structured and shaped by those that use them. For a sympathetic critique see Robinson 1994.

References

Abbott, A. 1988. *The System of Professions: An Essay on the Division of Expert Labor.* University of Chicago Press.

Adams, I. D., et al. 1986. Computer aided diagnosis of acute abdominal pain: A multicentre study. *British Medical Journal* 293: 800–804.

Agre, P. E. 1994. Design for Democracy. Unpublished manuscript, Department of Communication, University of California, San Diego.

Agre, P. E. 1995. From high tech to human tech: Empowerment, measurement, and social studies of computing. *Computer Supported Cooperative Work* 3: 167–195.

Aikins, J. S., J. C. Kunz, E. H. Shortliffe, and R. J. Fallat. 1983. PUFF: An expert system for interpetation of pulmonary function data. *Computers in Biomedical Research* 16: 199–208.

Akrich, M. 1992. The de-scription of technical objects. In *Shaping Technology/Building Society*, ed. W. Bijker and J. Law. MIT Press.

Akrich, M., and B. Latour. 1992. A summary of a convenient vocabulary for the semiotics of human and non-human assemblies. In *Shaping Technology/Building Society*, ed. W. Bijker and J. Law. MIT Press.

Allison, G. T. 1971. *Essence of Decision: Explaining the Cuban Missile Crisis.* Little, Brown.

Alvarez, W. 1953. On disregarding findings that cannot explain the syndrome. *New England Journal of Medicine* 249: 184–186.

Amann, K., and K. Knorr-Cetina. 1989. Thinking through talk: An ethnographic study of a molecular biology laboratory. *Knowledge and Society* 8: 3–26.

American Board of Internal Medicine. 1979. Clinical competence in internal medicine. *Annals of Internal Medicine* 90: 402–411.

Anderson, J. G., and S. J. Jay, eds. 1987. *Use and Impact of Computers in Clinical Medicine.* Springer-Verlag.

Anderson, W. 1992. The reasoning of the strongest: The polemics of skill and science in medical diagnosis. *Social Studies of Science* 22: 653–684.

Anonymous. 1922. Medical English as she is wrote. *Journal of the American Medical Association* 78: 731–732.

Anonymous. 1926. Filling the physician's prescription. *Journal of the American Medical Association* 87: 35–36.

Anonymous. 1938. Master in the house of medicine. *Journal of the American Medical Association* 111: 327.

Anonymous. 1945. A.M.A. program. *New England Journal of Medicine* 233: 223.

Anonymous. 1946. The inadequacies of medical care. I. General considerations. *New England Journal of Medicine* 234: 515–516.

Anonymous. 1947a. Basic science instruction in hospital residencies. *Journal of the American Medical Association* 133: 697–698.

Anonymous. 1947b. Diagnostic medical care and commercial laboratories. *Journal of the American Medical Association* 134: 457–458.

Anonymous. 1948a. Unnecessary operations. *New England Journal of Medicine* 238: 339–340.

Anonymous. 1948b. Congress looks at health. *New England Journal of Medicine* 239: 102.

Anonymous. 1948c. The *Providence Evening Bulletin* looks at medicine. *New England Journal of Medicine* 239: 208–209.

Anonymous. 1948d. The public wants a doctor when they want him. *Journal of the American Medical Association* 136: 695–696.

Anonymous. 1950a. Enrolments in medical schools. *Journal of the American Medical Association* 142: 420.

Anonymous. 1950b. The next fifty years. *Journal of the American Medical Association* 142: 34–35.

Anonymous. 1950c. Quality of medical care. *New England Journal of Medicine* 242: 381–382.

Anonymous. 1950d. Roadblocks to clinical research. *New England Journal of Medicine* 243: 677.

Anonymous. 1951a. "C.P.C.": Clinical analysis or guessing game? *New England Journal of Medicine* 245: 829–830.

Anonymous. 1951b. Physician-patient relationship. *Journal of the American Medical Association* 147: 1054.

Anonymous. 1952a. Quality of medical care. *New England Journal of Medicine* 247: 34–35.

Anonymous. 1952b. Clinical accountancy. *New England Journal of Medicine* 247: 142–143.

Anonymous. 1952c. The physician's responsibility. *Journal of the American Medical Association* 148: 471.

Anonymous. 1954a. A confusion of tongues. *Journal of the American Medical Association* 154: 1093.

Anonymous. 1954b. Uses of standard nomenclature. *Journal of the American Medical Association* 154: 586–587.

Anonymous. 1956a. Medical science and society. *New England Journal of Medicine* 254: 1190–1191.

Anonymous. 1956b. The epidemiologic method. *New England Journal of Medicine* 254: 1044–1045.

Anonymous. 1956c. Panorama of medical research. *Journal of the American Medical Association* 162: 209.

Anonymous. 1957a. Sequential probabilities. *New England Journal of Medicine* 256: 524.

Anonymous. 1957b. Standard nomenclature of disease and operations. *Journal of the American Medical Association* 163: 552.

Anonymous. 1958. From guesswork to guideline. *Journal of the American Medical Association* 166: 781.

Anonymous. 1959. President Eisenhower's Atlantic City address. *Journal of the American Medical Association* 170: 1072.

Anonymous. 1960. Medical research. *Journal of the American Medical Association* 173: 685.

Anonymous. 1961a. Computers and the practice of medicine. *Journal of the American Medical Association* 177: 205–206.

Anonymous. 1961b. Development of an industrial medical-records system. *Journal of the American Medical Association* 176: 805.

Anonymous. 1962. Doctors' dilemmas. *New England Journal of Medicine* 266: 1335.

Anonymous. 1963a. Electronic data processing apparatus. *Journal of the American Medical Association* 186: 146–147.

Anonymous. 1963b. Judgment difficult. *New England Journal of Medicine* 269: 1383–1384.

Anonymous. 1964. Physical diagnosis. *New England Journal of Medicine* 270: 476–477.

Anonymous. 1965a. Physical diagnosis: Requiem or renaissance? *Journal of the American Medical Association* 192: 164.

Anonymous. 1965b. Rending the veil of mystery. *Journal of the American Medical Association* 192: 128.

Anonymous. 1967. Report of the Bethesda Conferences of the Committee on Standardized Terminology of the American College of Cardiology. Glossary of cardiologic terms related to physical diagnosis and history. I. Heart murmurs. *Journal of the American Medical Association* 200: 1041–1042.

Anonymous. 1982. Algorithms for the clinician. *Lancet*, March 15: 528–529.

Anspach, R. R. 1993. *Deciding Who Lives: Fateful Choices in the Intensive Care Nursery.* University of California Press.

Aring, C. D. 1970. The patient's welfare and the medical record. *Journal of the American Medical Association* 214: 1317–1319.

Arkes, H. R., and K. R. Hammond, eds. 1986. *Judgment and Decision Making.* Cambridge University Press.

Armstrong, D. 1983. *Political Anatomy of the Body: Medical Knowledge in Britain in the Twentieth Century.* Cambridge University Press.

Arney, W., and B. Bergen. 1984. *Medicine and the Management of Living: Taming the Last Great Beast.* University of Chicago Press.

Ashmore, M., M. Mulkay, and T. Pinch. 1989. *Health and Efficiency: A Sociology of Health Economics.* Open University Press.

Ashmore, M., R. Wooffitt, and S. Harding. 1994. *Humans and Others, Agents and Things. American Behavioral Scientist* 37, no. 6 (special issue).

Atkinson, P. 1981. *The Clinical Experience: The Construction of Medical Reality.* Gower.

Atkinson, P. 1995. *Medical Talk and Medical Work.* Sage.

Audet, A. M., S. Greenfield, and M. Field. 1990. Medical practice guidelines: Current activities and future directions. *Annals of Internal Medicine* 113: 709–714.

Baars, B. J. 1986. *The Cognitive Revolution in Psychology.* Guilford.

Balla, J. I., A. S. Elstein, and C. Christensen. 1989. Obstacles to acceptance of clinical decision analysis. *British Medical Journal* 4: 579–582.

Barclay, W. 1978. Consensus development conferences. *Journal of the American Medical Association* 240: 378–379.

Barley, S. R. 1988. The social construction of a machine: Ritual, superstition, magical thinking and other pragmatic responses to running a CT scanner. In *Biomedicine Examined*, ed. M. Lock and D. Gordon. Kluwer.

Barnett, G. 1968. Computers in patient care. *New England Journal of Medicine* 279: 1321–1327.

Barnett, G. O. 1982. The computer and clinical judgement. *New England Journal of Medicine* 307: 493–494.

Barnett, G. O., J. J. Cimono, J. A. Hupp, and E. P. Hoffer. 1987. DXplain: An evolving diagnostic decision support system. *Journal of the American Medical Association* 258: 67–74.

Beecher, H. K. 1953. Clinical impression and clinical investigation. *Journal of the American Medical Association* 151: 44–45.

Bell, D. E., H. Raiffa, and A. Tversky. 1988. Descriptive, normative and prescriptive interactions in decision making. In *Decision Making*, ed. D. Bell et al. Cambridge University Press.

Beniger, J. R. 1986. *The Control Revolution: Technological and Economic Origins of the Information Society*. Harvard University Press.

Berg, M. 1992. The construction of medical disposals. medical sociology and medical problem solving in clinical practice. *Sociology of Health and Illness* 14: 151–180.

Berg, M. 1994. "Ze zijn allemaal dood . . .": Over kennissystemen in de medische praktijk, voorvechters en critici, en het mysterieuze succes van onderzoeksprotocollen. *Kennis en Methode* 18: 361–391.

Berg, M. 1996. Practices of reading and writing: The constitutive role of the patient record in medical work. *Sociology of Health and Illness* 18: 499–524.

Berg, M., and M. Casper, eds. 1995. *Constructivist Perspectives on Medical Practices*. Science, Technology, and Human Values 20, no. 3 (special issue).

Berg, M., and A. Mol, eds. Forthcoming. *Differences in Medicine: Unraveling Practices, Techniques and Bodies*.

Berwick, D. M. 1988. The society for medical decision making: The right place at the right time. *Medical Decision Making* 8: 77–80.

Bijker, W. E., and J. Law, eds. 1992. *Shaping Technology/Building Society: Studies in Sociotechnical Change*. MIT Press.

Bishop, L. F. 1900. A plea for greater simplicity in therapeutics. *Journal of the American Medical Association* 35: 1332–1333.

Bjerregaard, B., et al. 1976. Computer-aided diagnosis of the acute abdomen: A system from Leeds used on Copenhagen patients. In *Decision Making and Medical Care*, ed. F. de Dombal and F. Gremy. North-Holland.

Bleich, H. L. 1971. Prognosis by calculation. *New England Journal of Medicine* 285: 1533–1534.

Bleich, H. L. 1972. Computer-based consultation: Electrolyte and acid-base disorders. *American Journal of Medicine* 53: 285–291.

Blois, M. S. 1980. Clinical judgment and computers. *New England Journal of Medicine* 303: 192–197.

Bloomfield, B. P. 1991. The role of information systems in the UK national health service: Action at a distance and the fetish of calculation. *Social Studies of Science* 21: 701–734.

Bloor, M. 1976. Bishop Berkeley and the adeno-tonsillectomy enigma. *Sociology* 10: 43–61.

Bloor, M. 1978. On the routinised nature of work in people-processing agencies: The case of adenotonsillectomy assessments in ENT out-patient clinics. In *Relationships between Doctors and Patients*, ed. A. Davis. Gower.

Blume, S. S. 1992. *Insight and Industry: On the Dynamics of Technological Change in Medicine*. MIT Press.

Böckenholt, U., and E. U. Weber. 1992. Use of formal methods in medical decision making. *Medical Decision Making* 12: 298–306.

Boon, L. 1982. *Geschiedenis van de psychologie*. Boom.

Borak, J., and S. Veilleux. 1982. Errors of intuitive logic among physicians. *Social Science and Medicine* 16: 1939–1942.

Bosk, C. L. 1979. *Forgive and Remember: Managing Medical Failure*. University of Chicago Press.

Bowers, J. 1992. The politics of formalism. In *Contexts of Computer Mediated Communication*, ed. M. Lea. Harvester.

Bowker, G. 1994a. Dismembering and remembering: Classification and organizational memory. Paper presented at Conference on Locating Design, Development and Use, Oksnoen.

Bowker, G. 1994b. *Science on the Run: Information Management and Industrial Geophysics at Schlumberger, 1920–1940*. MIT Press.

Bowker, G., and S. L. Star. 1994. Knowledge and infrastructure in international information management: Problems of classification and coding. In *Information Acumen*, ed. L. Bud. Routledge.

Braybrooke, D., and C. E. Lindblom. 1963. *A Strategy of Decision: Policy Evaluation as a Social Process*. Free Press.

Brook, R. H., et al. 1988. Diagnosis and treatment of coronary disease: comparison of doctors' attitudes in the USA and the UK. *Lancet* April 2: 750–753.

Buchanan, B. G., and E. H. Shortliffe, eds. 1984. *Rule-Based Expert Systems: The MYCIN Experiments of the Stanford Heuristic Programming Project*. Addison-Wesley.

Bursztajn, H. et al. 1981. *Medical Choices, Medical Chances: How Patients, Families and Physicians Can Cope with Uncertainty*. Dell.

Button, G., ed. 1993. *Technology in Working Order: Studies of Work, Interaction, and Technology*. Routledge.

Callon, M. 1980. Struggles and negotiations to decide what is problematic and what is not: The sociology of translation. In *The Social Process of Scientific Investigation*, ed. K. Knorr et al. Reidel.

Callon, M. 1987. Society in the making: The study of technology as a tool for sociological analysis. In *The Social Construction of Technological Systems*, ed. W. Bijker et al. MIT Press.

Callon, M. 1991. Techno-economic networks and irreversibility. In *A Sociology of Monsters*, ed. J. Law. Routledge.

Callon, M., and J. Law. 1995. Agency and the hybrid *Collectif*. *South Atlantic Quarterly* 94: 481–507.

Cambrosio, A., and P. Keating. 1992. A Matter of FACS: Constituting novel entities in immunology. *Medical Anthropology Quarterly* 6: 362–384.

Casparie, A. F., and J. J. E. Everdingen. 1989. Maatstaven voor goede zorg: Vijf jaar later. *Medisch Contact* 44: 1654–1656.

Cebul, R. D. 1984. A look at the chief complaints revisited: Current obstacles and opportunities for decision analysis. *Medical Decision Making* 4: 271–283.

Cebul, R. D. 1988. Decision making research at the interface between descriptive and prescriptive studies. *Medical Decision Making* 8: 231–232.

Chatbanchai, W. et al. 1989. Acute abdominal pain and appendicitis in North East Thailand. *Paediatric and Perinatal Epidemiology* 3: 448–459.

Cicourel, A. V. 1986. The reproduction of objective knowledge: Common sense reasoning in medical decision making. In *The Knowledge Society*, ed. G. Böhme and N. Stehr. Reidel.

Clancey, W. J., and E. H. Shortliffe, eds. 1984. *Readings in Medical Artificial Intelligence: The First Decade*. Addison-Wesley.

Clancy, C. M., R. D. Cebul, and S. V. Williams. 1988. Guiding individual decisions: A randomized, controlled trial of decision analysis. *American Journal of Medicine* 84: 283–288.

Clarke, A. E., and J. H. Fujimura, eds. 1992. *The Right Tools for the Job: At Work in Twentieth-Century Life Sciences*. Princeton University Press.

Clarke, J. R. 1989. Appendicitis: The computer as a diagnostic tool. *International Journal of Technology Assessment in Health Care* 5: 371–379.

Clinton, J. J. 1992. Improving clinical practice. *Journal of the American Medical Association* 267: 2652–2653.

Clouser, K. D. 1985. Approaching the logic of diagnosis. In *Logic of Discovery and Diagnosis in Medicine*, ed. K. Schaffner. University of California Press.

Colcock, B. 1974. PSRO's. *New England Journal of Medicine* 290: 1318–1319.

Collings, J. S., and D. M. Clark. 1953. General practice, today and tomorrow. *New England Journal of Medicine* 248: 141–148, 183–194.

Collins, H. M. 1985. *Changing Order: Replication and Induction in Scientific Practice.* Sage.

Collins, H. M. 1990. *Artificial Experts: Social Knowledge and Intelligent Machines.* MIT Press.

Cornfield, J. 1972. Statistical classification methods. In *Computer Diagnosis and Diagnostic Methods*, ed. J. Jacquez. Thomas.

Coulter, J. 1983. *Rethinking Cognitive Theory.* St. Martin's.

Coulter, J. 1989. *Mind in Action.* Polity Press.

Crosby, W. H. 1977. Chess and combat: The algorithm in medicine. *Journal of the American Medical Association* 238: 2721.

Cummins, R. O. 1990. Decision analysis, the journal of general internal medicine, and the general internist. *Journal of General Internal Medicine* 5: 375–378.

Cussins, C. Forthcoming. Ontological choreography: Agency for women patients in an infertility clinic. In *Differences in Medicine*, ed. M. Berg and A. Mol.

Cutler, P. 1979. *Problem Solving in Clinical Medicine.* Williams and Wilkins. (Cited in Anonymous 1982.)

Danilevicius, Z. 1975. The power of unbiased observation. *Journal of the American Medical Association* 231: 966.

Davis, K. 1986. The process of problem (re)formulation in psychotherapy. *Sociology of Health and Illness* 8: 44–74.

Davis, R., B. G. Buchanan, and E. H. Shortliffe. 1977. Production rules as a representation for a knowledge-based consultation program. *Artificial Intelligence* 8: 15–45.

de Dombal, F. T. 1987a. Back to the future; or forward to the past? *Gut* 28: 373–377.

de Dombal, F. T. 1987b. Computer-aided decision support: The obstacles to progress. *Methods of Information in Medicine* 26: 183–184.

de Dombal, F. T. 1988. The OMGE acute abdominal pain survey: Progress report, 1986. *Scandinavian Journal of Gastroenterology* 23 (supplement 144): 35–42.

de Dombal, F. T. 1989. Computer-aided decision support in clinical medicine. *International Journal of Biomedicine and Computing* 24: 9–16.

de Dombal, F. T. 1990. Computer-aided decision support: Glittering prospects, practical problems, and Pandora's Box. *Baillière's Clinical Obstetrics and Gynaecology* 4: 841–849.

de Dombal, F. T. 1991. The diagnosis of acute abdominal pain with computer assistance: Worldwide perspective. *Annales de Chirurgie* 45: 273–277.

de Dombal, F. T., V. Dallos, and W. A. McAdam. 1991. Can computer aided teaching packages improve clinical care in patients with acute abdominal pain? *British Medical Journal* 302: 1495–1497.

de Dombal, F. T., and F. Gremy, eds. 1976. *Decision Making and Medical Care.* North-Holland.

de Dombal, F. T., D. J. Leaper, J. R. Staniland, A. P. McCann, and J. C. Horrocks. 1972. Computer-aided diagnosis of acute abdominal pain. *British Medical Journal* 2: 9–13.

de Vries, G. 1993. *Gerede Twijfel: Over de Rol van de Medische Ethiek in Nederland.* De Balie.

de Vries, P. H., and P. F. de Vries-Robbe. 1985. An overview of medical expert systems. *Methods of Information in Medicine* 24: 57–64.

Dent, M. 1990. Organisation and change in renal work: a study of the impact of a computer system within two hospitals. *Sociology of Health and Illness* 12: 413–431.

Detsky, A. S. 1987. Decision analysis: What's the prognosis? *Annals of Internal Medicine* 106: 321–322.

Detsky, A. S., D. Redelmeier, and H. B. Abrams. 1987. What's wrong with decision analysis? Can the left brain influence the right? *Journal of Chronic Diseases* 40: 831–838.

Dodier, N. 1995. *Les hommes et les machines.* Métailié.

Dodier, N. Forthcoming. clinical practice and procedures in occupational medicine: A study of the framing of individuals. In *Differences in Medicine*, ed. M. Berg and A. Mol.

Dolan, J. G. 1990. Can decision analysis adequately represent clinical problems? *Journal of Clinical Epidemiology* 43: 277–284.

Doubilet, P., and B. J. McNeil. 1985. Clinical decisionmaking. *Medical Care* 23: 648–662.

Doyle, J. C. 1952. Unnecessary ovariectomies: Study based on removal of 704 normal ovaries from 546 patients. *Journal of the American Medical Association* 148: 1105–1111.

Dreyfus, H. 1972. *What Computers Can't Do: A Critique of Artificial Reason.* See *What Computers Still Can't Do* (MIT Press, 1992).

Dreyfus, H., and S. E. Dreyfus. 1986. *Mind over Machine: The Power of Human Intuition and Expertise in the Era of the Computer.* Blackwell.

Duda, R. O., and E. H. Shortliffe. 1983. Expert systems research. *Science* 220: 261–268.

Eagle, K. E. 1991. Medical decision making in patients with chest pain. *New England Journal of Medicine* 324: 1282–1283.

Ebbesen, E. B., and V. J. Konecni. 1980. On the external validity of decision-making research: What do we know about decisions in the real world? In *Cognitive Processes, Choice and Decision Behavior*, ed. T. Wallsten. Erlbaum.

Eddy, D. M. 1982. Probabilistic reasoning in clinical medicine: Problems and opportunities. In *Judgment under Uncertainty*, ed. D. Kahneman et al. Cambridge University Press.

Eddy, D. M. 1990a. Anatomy of a decision. *Journal of the American Medical Association* 263: 441–443.

Eddy, D. M. 1990b. The challenge. *Journal of the American Medical Association* 263: 287–290.

Eddy, D. M. 1990c. Practice policies: What are they? *Journal of the American Medical Association* 263: 877–880.

Eddy, D. M. 1990d. Practice Policies: Where do they come from? *Journal of the American Medical Association* 263: 1265–1275.

Eddy, D. M. 1990e. Practice policies: Guidelines for methods. *Journal of the American Medical Association* 263: 1839–1841.

Eddy, D. M. 1990f. Guidelines for policy statements: The explicit approach. *Journal of the American Medical Association* 263: 2239–2243.

Eddy, D. M. 1992. Applying cost-effectiveness analysis: The inside story. *Journal of the American Medical Association* 268: 2575–2582.

Edwards, D. 1994. Imitations and artifice in apes, humans, and machines. *American Behavioral Scientist* 37: 754–771.

Edwards, P. N. 1996. *The Closed World: Computers and the Politics of Discourse in Cold War America*. MIT Press.

Eisenberg, J. M. 1986. *Doctors' Decisions and the Cost of Medical Care*. Health Administration Press.

Elstein, A. S. 1982. Comment [on Borak and Veilleux 1982]. *Social Science and Medicine* 16: 1945–1946.

Elstein, A. S. et al. 1986. Comparison of physicians' decisions regarding estrogen replacement therapy for menopausal women and decisions derived from a decision analytic model. *American Journal of Medicine* 80: 246–258.

Elstein, A. S., L. S. Schulman, and S. A. Sprafka. 1978. *Medical Problem Solving*. Harvard University Press.

Elstein, A. S., L. S. Shulman, and S. A. Sprafka. 1990. Medical problem solving: a ten-year retrospective. *Evaluation and the Health Professions Special Issue: Reflections on Research in Medical Problem Solving* 13: 5–36.

Emerson, P. A., J. Wyatt, L. Dillistone, N. Crichton, and N. J. Russell. 1988. The development of ACORN, an expert system enabling nurses to make admission decisions about patients with chest pain in an accident and emergency department. In *Proceedings of Medical Informatics: Computers in Clinical Medicine*. British Medical Informatics Society.

Engle, R. L. 1992. Attempts to use computers as diagnostic aids in medical decision making: a thirty-year experience. *Perspectives in Biology and Medicine* 35: 207–219.

Epstein, S. 1995. The critique of pure science: AIDS activism and the struggle for credibility in the reform of clinical trials. In *Constructivist Perspectives on Medical Practices*, ed. M. Casper and M. Berg (*Science, Technology, and Human Values* 20, no. 3).

Evans, R. G. 1990. The dog in the night-time: medical practice variations and health policy. In *The Challenges of Medical Practice Variations*, ed. T. Anderson and G. Mooney. Macmillan.

Everdingen, J. J. E. 1988. *Consensusontwikkeling in de geneeskunde*. Bohn, Scheltema en Holkema.

Feinstein, A. R. 1967. *Clinical Judgment*. Krieger.

Feinstein, A. R. 1973a. An analysis of diagnostic reasoning. I. The domains and disorders of clinical macrobiology. *Yale Journal of Biology and Medicine* 46: 212–232.

Feinstein, A. R. 1973b. An analysis of diagnostic reasoning. II. The strategy of intermediate decisions. *Yale Journal of Biology and Medicine* 46: 264–283.

Feinstein, A. R. 1973c. The problems of the "problem-oriented medical record." *Annals of Internal Medicine* 78: 751–762.

Feinstein, A. R. 1974. An analysis of diagnostic reasoning. III. The construction of clinical algorithms. *Yale Journal of Biology and Medicine* 1: 5–32.

Feinstein, A. R. 1977. Clinical biostatistics xxxix: The haze of Bayes, the aerial palaces of decision analysis, and the computerized Ouija board. *Clinical Pharmacology and Therapeutics* 21: 482–496.

Feinstein, A. R. 1979. Science, clinical medicine, and the spectrum of disease. In *Cecil Textbook of Medicine*, ed. P. Beeson et al. Saunders.

Feinstein, A. R. 1985. *Clinical Epidemiology: The Architecture of Clinical Research*. Saunders.

Feinstein, A. R. 1987a. *Clinimetrics*. Yale University Press.

Feinstein, A. R. 1987b. The intellectual crisis in clinical science: Medaled models and muddled mettle. *Perspectives in Biology and Medicine* 30: 215–230.

Feinstein, A. R. 1994. Clinical judgment revisited: The distraction of quantitative models. *Annals of Internal Medicine* 120: 799–805.

Fenyo, G. et al. 1987. Computer-aided diagnosis of 233 acute abdominal cases at Nacka Hospital Sweden. *Scandinavian Journal of Gastroenterology* Supplement 128: 178.

Field, M. J., and K. N. Lohr, eds. 1990. *Clinical Practice Guidelines: Directions for a New Program.* National Academy Press.

Fischoff, B. 1988. Clinical decision making. In *Professional Judgment*, ed. J. Dowie and A. Elstein. Cambridge University Press.

Fisher, S., and A. D. Todd, eds. 1983. *The Social Organization of Doctor-Patient Communication.* Center for Applied Linguistics.

Flanagin, A., and G. D. Lundberg. 1990. Clinical decision making: Promoting the jump from theory to practice. *Journal of the American Medical Association* 263: 279–280.

Flexner, A. 1910. *Medical Education in the United States and Canada.* Carnegie Foundation for the Advancement of Teaching.

Forsythe, D. E. 1992. Blaming the user in medical informatics: The cultural nature of scientific practice. *Knowledge and Society: The Anthropology of Science and Technology* 9: 95–111.

Forsythe, D. E. 1993a. The construction of work in artificial intelligence. *Science, Technology, and Human Values* 18: 460–479.

Forsythe, D. E. 1993b. Engineering knowledge: The construction of knowledge in artificial intelligence. *Social Studies of Science* 23: 445–477.

Frankenberg, R., ed. 1992. *Time, Health and Medicine.* Sage.

Friedman, A., and D. S. Comford. 1989. *Computer Systems Development: History, Organization and Implementation.* Wiley.

Frohock, F. M. 1986. *Special Care: Medical Decisions at the Beginning of Life.* University of Chicago Press.

Froom, J. 1975. International classification of health problems in primary care. *Journal of the American Medical Association* 234: 1257.

Fujimura, J. H. 1987. Constructing "do-able" problems in cancer-research: Articulating alignment. *Social Studies of Science* 17: 257–293.

Fujimura, J. H. 1988. The molecular biological bandwagon in cancer research: Where social worlds meet. *Social Problems* 35: 261–283.

Fujimura, J. H. 1992. Crafting science: Standardized packages, boundary objects, and "translation." In *Science as Practice and Culture*, ed. A. Pickering. University of Chicago Press.

Fuller, S. 1991. Harry the Apostate (or PC-squared) (Response to Collins's *Artificial Experts*). Paper presented at *Annual Meeting of the Society of the Social Studies of Science*, Boston, November 1991.

Fuller, S. 1993a. *Philosophy, Rhetoric, and the End of Knowledge: The Coming of Science and Technology Studies*. University of Wisconsin Press.

Fuller, S. 1993b. What Dreyfus still can't see [review of Dreyfus's *What Computers Still Can't Do*]. *EAST-Newsletter* 12: 11–15.

Garfinkel, H. 1967. *Studies in Ethnomethodology*. Prentice-Hall.

Garland, L. H. 1959. Studies on the accuracy of diagnostic procedures. *American Journal of Roentgenology* 82: 25–38.

Gasser, L. 1986. The integration of computing and routine work. *ACM Transactions on Office Information Systems* 4: 205–225.

Gigerenzer, G., and D. J. Murray. 1987. *Cognition as Intuitive Statistics*. Erlbaum.

Gigerenzer, G. et al. 1989. *The Empire of Chance: How Probability Changed Science and Everyday Life*. Cambridge University Press.

Gill, P. W. et al. 1973. Observer variation in clinical diagnosis: A computer-aided assessment of its magnitude and importance in 552 patients with abdominal pain. *Methods of Information in Medicine* 12: 108–113.

Ginsberg, A. S. 1972. The diagnostic process viewed as a decision problem. In *Computer Diagnosis and Diagnostic Methods*, ed. J. Jacquez. Thomas.

Goldwater, L. J. 1959. Clinical judgment and common sense for the cardiac patient. *Journal of the American Medical Association* 169: 598.

Goldwater, L. J., L. H. Bronstein, and B. Kresky. 1952. Study of one hundred seventy-five "cardiacs" without heart disease. *Journal of the American Medical Association* 148: 89–92.

Good, M. 1995. *American Medicine: The Quest for Competence*. University of California Press.

Gooding, D. 1992. Putting agency back into experiment. In *Science as Practice and Culture*, ed. A. Pickering. University of Chicago Press.

Goody, J. 1977. *The Domestication of the Savage Mind*. Cambridge University Press.

Gordon, D. R. 1988. Clinical science and expertise: Changing boundaries between art and science in medicine. In *Biomedicine Examined*, ed. M. Lock and D. Gordon. Kluwer.

Gorry, G. A., J. P. Kassirer, A. Essig, and W. B. Schwartz. 1973. Decision analysis as the basis for computer-aided management of acute renal failure. *American Journal of Medicine* 55: 473–484.

Graves, W. III, and J. M. Nyce. 1992. Normative models and situated practice in medicine: Towards more adequate system design and development. *Information and Decision Technologies* 19: 143–149.

Greenbaum, J., and M. Kyng, eds. 1991. *Design at Work: Cooperative Design for Computer Systems.* Erlbaum.

Greenberg, R. N. 1978. An argument for research in the medical school curriculum. *Journal of the American Medical Association* 239: 1162–1163.

Greenfield, S., F. E. Bragg, D. L. McCraith, and J. Blackburn. 1974. Upper-respiratory tract complaint protocol for physician-extenders. *Archives of Internal Medicine* 133: 294–299.

Grimm, R. H., K. Shimoni, W. R. Harlan, and E. H. Estes. 1975. Evaluation of patient-care protocol use by various providers. *New England Journal of Medicine* 292: 507–511.

Groen, G. J., and V. L. Patel. 1985. Medical problem-solving: some questionable assumptions. *Medical Education* 19: 95–100.

Grol, R. 1989. De verspreiding van NHG standaarden onder huisartsen. *Huisarts en Wetenschap* 32: 494–497.

Gustafson, D. H., R. L. Ludke, P. J. Glackman, F. C. Larson, and J. H. Greist. 1972. Wisconsin Computer Aided Medical Diagnosis Project: Progress report. In *Computer Diagnosis and Diagnostic Methods*, ed. J. Jacquez. Thomas.

Haggerty, R. 1973. Effectiveness of medical care. *New England Journal of Medicine* 289: 372–373.

Hanseth, O., E. Monteiro, and M. Hatling. Forthcoming. Developing information infrastructure: The tension between standardisation and flexibility. *Science, Technology and Human Values.*

Haraway, D. J. 1991. *Simians, Cyborgs, and Women: The Reinvention of Nature.* Routledge.

Haraway, D. J. 1994. A game of cat's cradle: Science studies, feminist theory, cultural studies. *Configurations* 1: 59–71.

Harper, R. H. R., and J. A. Hughes. 1993. "What a F-ing System! Send 'em all to the same place and then expect us to stop 'em hitting": Making technology work in air traffic control. In *Technology in Working Order*, ed. G. Button. Routledge.

Hart, A., and J. Wyatt. 1989. Connectionist models in medicine: An investigation of their potential. In *Proceedings of AIME 89. Second European Conference on Artificial Intelligence in Medicine*, ed. J. Hunter et al. Springer-Verlag.

Hartland, J. 1993a. The Machinery of Medicine: An Analysis of Algorithmic Approaches to Medical Knowledge and Practice. Ph.D. thesis, University of Bath.

Hartland, J. 1993b. The use of "intelligent" machines for electrocardiograph interpretation. In *Technology in Working Order*, ed. G. Button. Routledge.

Haugeland, J. 1985. *Artificial Intelligence: The Very Idea.* MIT Press.

Hayes-Roth, F., D. A. Waterman, and D. B. Lenat. 1983. *Building Expert Systems.* Addison-Wesley.

Hayles, N. K. 1994. Boundary disputes: Homeostasis, reflexivity, and the foundations of cybernetics. *Configurations* 3: 441–467.

Haynes, D., et al. 1986. How to keep up with the medical literature: How to store and retrieve articles worth keeping. *Annals of Internal Medicine* 105: 978–984. Quoted in Wyatt 1991b.

Heathfield, H. A., and J. Wyatt. 1993. Philosophies for the design and development of clinical decision-support systems. *Methods of Information in Medicine* 32: 1–8.

Heims, S. J. 1980. *John von Neumann and Norbert Wiener: From Mathematics to the Technologies of Life and Death.* MIT Press.

Helman, C. G. 1988. Psyche, soma and society: The social construction of psychosomatic disorders. In *Biomedicine Examined,* ed. M. Lock and D. Gordon. Kluwer.

Heritage, J. 1984. *Garfinkel and Ethnomethodology.* Polity Press.

Hershey, J. C., and J. Baron. 1987. Clinical reasoning and cognitive processes. *Medical Decision Making* 7: 203–211.

Hill, M. N., and C. S. Weisman. 1991. Physicians' perceptions of consensus reports. *International Journal of Technology Assessment in Health Care* 7: 30–41.

Hirschauer, S. 1991. The manufacture of bodies in surgery. *Social Studies of Science* 21: 217–319.

Hirschauer, S. Forthcoming. Shifting contradictions. doing sex and doing gender in medical disciplines. In *Differences in Medicine,* ed. M. Berg and A. Mol.

Holmes-Rovner, M. 1992. Methods for medical decision making. Presidential Address. *Medical Decision Making* 12: 159–162.

Horrocks, J. C., W. A. F. McAdam, G. Devroede, A. A. Gunn, and N. Zoltie. 1976. Some practical problems in transferring computer-aided diagnostic systems from one geographical area to another. In *Decision Making and Medical Care,* ed. F. de Dombal and F. Gremy. North-Holland.

Horrocks, J. C., A. P. McCann, J. R. Staniland, D. J. Leaper, and F. T. de Dombal. 1972. Computer-aided diagnosis: Description of an adaptable system, and operational experience with 2,034 cases. *British Medical Journal* 2: 5–9.

Horstman, K. Forthcoming. Chemical analysis of urine for life insurance: The construction of reliability. *Science, Technology, and Human Values.*

Howard, J. M. 1961. Cooperative clinical research programs between medical institutions. *Journal of the American Medical Association* 175: 705–706.

Hubbard, W. N. 1964. Welcome and prologue. In *The Diagnostic Process*, ed. J. Jacquez. Mallory.

Hughes, D. 1988. When nurse knows best: Some aspects of nurse/doctor interaction in a casualty department. *Sociology of Health and Illness* 10: 1–22.

Hunter, K. M. 1991. *Doctor's Stories: The Narrative Structure of Medical Knowledge.* Princeton University Press.

Hurst, J. W. 1971. Ten reasons why Lawrence Weed is right. *New England Journal of Medicine* 284: 51–52.

Hutchins, E. 1995. *Cognition in the Wild.* MIT Press.

Ikonen, J. K. et al. 1983. Presentation and diagnosis of acute abdominal pain in Finland: A computer aided study. *Annales Chirurgiae et Gynaecologiae* 72: 332–336.

Ingelfinger, F. J. 1973. Algorithms, anyone? *New England Journal of Medicine* 288: 847–848.

Ingelfinger, F. J. 1975. Decision in medicine. *New England Journal of Medicine* 293: 254–255.

Jacoby, I. 1988. Evidence and consensus. *Journal of the American Medical Association* 259: 3039.

Jacoby, I., and S. G. Pauker. 1986. Technology assessment in health care: Group process and decision theory. *Israel Journal of Medical Sciences* 22: 183–190.

Jacquez, J. A. 1964. *The Diagnostic Process.* Mallory.

Jacquez, J. A. 1972. *Computer Diagnosis and Diagnostic Methods.* Thomas.

Jencks, S. F. 1992. Accuracy in recorded diagnoses. *Journal of the American Medical Association* 267: 2238–2239.

Jordan, K., and M. Lynch. 1992. The sociology of a genetic engineering technique: Ritual and rationality in the performance of the "plasmid prep." In *The Right Tools for the Job*, ed. A. Clark and J. Fujimura. Princeton University Press.

Jordan, K., and M. Lynch. 1993. The mainstreaming of a molecular biological tool: A case study of a new technique. In *Technology in Working Order*, ed. G. Button. Routledge.

Judd, W. H. 1960. Physicians in a changing world. *New England Journal of Medicine* 263: 894–899.

Kahan, J. P., D. E. Kanouse, and J. D. Winkler. 1988. Stylistic variations in national institutes of health consensus statements, 1979–1983. *International Journal of Technology Assessment in Health Care* 4: 289–304.

Kahneman, D., P. Slovic, and A. Tversky. 1982. *Judgment under Uncertainty: Heuristics and Biases.* Cambridge University Press.

Kanouse, D. E., and I. Jacoby. 1988. When does information change practitioners' behavior? *International Journal of Technology Assessment in Health Care* 4: 27–33.

Kanouse, D. E. et al. 1989. *Changing Medical Practice through Technology Assessment: An Evaluation of the NIH Consensus Development Program.* Health Administration Press.

Kaplan, B. 1987. The medical computing "lag": Perceptions of barriers to the application of computers to medicine. *International Journal of Technology Assessment in Health Care* 3: 123.

Kaplan, B. 1995. The computer prescription: Medical computing, public policy, and views of history. *Science, Technology, and Human Values* 20: 5–38.

Kassirer, J. P., and R. I. Kopelman. 1988. Intuitive and inspirational, or inductive and incremental? *Hospital Practice* 23: 21–27.

Kassirer, J. P., and R. I. Kopelman. 1990. Diagnosis and decisions by algorithms. *Hospital Practice* 25: 23–31.

Kassirer, J. P., A. J. Moskowitz, J. Lau, and S. G. Pauker. 1987. Decision analysis: A progress report. *Annals of Internal Medicine* 106: 275–291.

Kassirer, J. P., and S. G. Pauker. 1978. Should diagnostic testing be regulated? *New England Journal of Medicine* 299: 947–949.

Kling, R. 1991. Computerization and social transformations. *Science, Technology, and Human Values* 16: 342–367.

Kling, R., and S. Iacono. 1984. The control of information systems developments after implementation. *Communications of the ACM* 27: 1218–1226.

Kling, R., and W. Scacchi. 1982. The web of computing: Computer technology as social organization. *Advances in Computers* 21: 1–90.

Knaus, W. A. 1986. Rationing, justice, and the American physician. *Journal of the American Medical Association* 255: 1176–1177.

Knaus, W. A., D. P. Wagner, and J. Lynn. 1991. Short-term mortality predictions for critically ill hospitalized adults: Science and ethics. *Science* 254: 389–394.

Knorr-Cetina, K., and K. Amann. 1990. Image dissection in natural scientific inquiry. *Science, Technology, and Human Values* 15: 259–283.

Knorr-Cetina, K. D. 1981. *The Manufacture of Knowledge.* Pergamon.

Knorr-Cetina, K. 1995. How superorganisms change: Consensus formation and the social ontology of high-energy physics experiments. *Social Studies of Science* 25: 119–147.

Knottnerus, J. 1987. Bouwstenen voor een rationele medische besluitvorming. *Medisch Contact* 42: 501–504.

Komaroff, A. L. 1982. Algorithms and the "art" of medicine. *American Journal of Public Health* 72: 10–12.

Komaroff, A. L. et al. 1974. Protocols for physician assistants: Management of diabetes and hypertension. *New England Journal of Medicine* 290: 307–312.

Komaroff, A. L., B. Reiffen, and H. Sherman. 1973. Problem-oriented protocols for physician-extenders. In *Applying the Problem-Oriented System*, ed. H. Walker et al. Medcom.

Koontz, A. R. 1959. The humanities in medicine. *Journal of the American Medical Association* 170: 72–73.

Korpman, R. A., and T. L. Lincoln. 1988. The computer-stored medical record: For whom? *Journal of the American Medical Association* 259: 3454–3456.

Kosecoff, J. et al. 1987. Effects of the National Institutes of Health Consensus Development Program on physician practice. *Journal of the American Medical Association* 258: 2708–2713.

Kulikowski, C. A. 1977. Problems in the design of knowledge bases for medical consultation. In *IEEE Proceedings of the First Annual Symposium on Computer Applications in Medical Care*. IEEE.

Langlotz, C. P. 1989. The feasibility of axiomatically-based expert systems. *Computer Methods and Programs in Biomedicine* 30: 85–95.

Latour, B. 1986. Visualisation and cognition: Thinking with eyes and hands. *Knowledge and Society* 6: 1–40.

Latour, B. 1987. *Science in Action*. Open University Press.

Latour, B. 1988. *The Pasteurization of France*. Harvard University Press.

Latour, B. 1993. *We Have Never Been Modern*. Harvard University Press.

Latour, B., and S. Woolgar. 1986. *Laboratory Life: The Construction of Scientific Facts*. Princeton University Press.

Lau, J. M., J. P. Kassirer, and S. G. Pauker. 1983. Decision Maker 3.0: Improved decision analysis by personal computer. *Medical Decision Making* 3: 39–43.

Lave, J. 1988. *Cognition in Practice*. Cambridge University Press.

Law, J. 1987. Technology and heterogeneous engineering: The case of portuguese expansion. In *The Social Construction of Technological Systems*, ed. W. Bijker et al. MIT Press.

Law, J. 1994. *Organizing Modernity*. Blackwell.

Law, J., and A. Mol. Forthcoming. On hidden heterogeneities: The design of an aircraft.

Leaper, D. J., J. C. Horrocks, J. R. Staniland, and F. T. de Dombal. 1972. Computer-assisted diagnosis of abdominal pain using "estimates" provided by clinicians. *British Medical Journal* 4: 350–354.

Ledley, R. S. 1965. *Use of Computers in Biology and Medicine*. McGraw-Hill.

Ledley, R. S., and L. B. Lusted. 1959. Reasoning foundations of medical diagnosis: Symbolic logic, probability, and value theory aid our understanding of how physicians reason. *Science* 130: 9–21.

Lee, N., and S. Brown. 1994. Otherness and the actor-network: The undiscovered continent. *American Behavioral Scientist* 37: 772–790.

Lipscombe, B. 1989. Expert systems and computer-controlled decision making in medicine. *AI and Society* 3: 184–197.

Lipscombe, B. 1991. Minds, Machines and Medicine: An Epistemological Study of Computer Diagnosis. Ph.D. thesis, University of Bath.

Lister, J. 1957. Time to think—about socialized medicine. *New England Journal of Medicine* 256: 221–222.

Littenberg, B., and H. C. Sox Jr. 1988. Evaluating individualized medical decision analysis. *American Journal of Medicine* 84: 289–290.

LoGerfo, J. P. 1977. Variation in surgical rates: Fact vs. fantasy. *New England Journal of Medicine* 297: 387–389.

Lomas, J. 1990. Promoting clinical policy change: Using the art to promote the science in medicine. In *The Challenges of Medical Practice Variations*, ed. T. F. Andersen and G. Mooney. Macmillan.

Lomas, J. 1993. Making clinical policy explicit: Legislative policy making and lessons for developing practice guidelines. *International Journal of Technology Assessment in Health Care* 9: 11–25.

Lomas, J. et al. 1988. The role of evidence in the consensus process: Results from a Canadian consensus exercise. *Journal of the American Medical Association* 259: 3001–3005.

Lomas, J. et al. 1989. Do practice guidelines guide practice? The effect of a consensus statement on the practice of physicians. *New England Journal of Medicine* 321: 1306–1311.

Lomas, J. A., et al. 1991. Opinion leaders vs. audit and feedback to implement practice guidelines. *Journal of the American Medical Association* 265: 2202–2207.

Löwy, I. Forthcoming. *Between Bench and Bedside: Science, Healing and Interleukin-2 in a Cancer Ward*. Harvard University Press.

Löwy, I. 1995. Nothing more to do: Palliative care versus experimental therapy in advanced cancer. *Science in Context* 8: 209–230.

Luff, P., and C. Heath. 1993. System use and social organisation: Observations on human-computer interaction in an architectural practice. In *Technology in Working Order*, ed. G. Button. Routledge.

Lundberg, G. D. 1981. Acting on significant laboratory results. *Journal of the American Medical Association* 245: 1762–1763.

Lundsgaarde, H. P. 1987. Evaluating medical expert systems. *Social Science and Medicine* 24: 805–819.

Lusted, L. B. 1968. *Introduction to Medical Decision Making*. Thomas.

Lusted, L. B. 1971. Decision-making studies in patient management. *New England Journal of Medicine* 284: 416–424.

Lusted, L. B. 1975. In the process of solution. *New England Journal of Medicine* 293: 255–256.

Lynch, M. 1984. "Turning up signs" in neurobehavioral diagnosis. *Symbolic Interaction* 7: 67–86.

Lynch, M. 1985. *Art and Artifact in Laboratory Science: A Study of Shop Work and Shop Talk in a Research Laboratory*. Routledge and Kegan Paul.

Lynch, M. 1991. Laboratory space and the technological complex: An investigation of topical contextures. *Science in Context* 4: 51–78.

Lynch, M. 1993. *Scientific Practice and Ordinary Action: Ethnomethodology and Social Studies of Science*. Cambridge University Press.

Lynch, M., E. Livingston, and H. Garfinkel. 1983. Temporal order in laboratory work. In *Science Observed*, ed. K. Knorr-Cetina and M. Mulkay. Sage.

MacMahon, B. 1955. Statistical methods in medicine. *New England Journal of Medicine* 253: 646–652, 688–693.

Margolis, C. Z. 1983. Uses of clinical algorithms. *Journal of the American Medical Association* 249: 627–632.

Marks, H. M. 1988. Notes from the underground: The social organization of therapeutic research. In *Grand Rounds*, ed. R. Maulitz and D. Long. University of Pennsylvania Press.

Marks, H. M. 1997. *The Progress of Experiment: Science and Therapeutic Reform in the United States, 1900–1990*. Cambridge University Press.

Markussen, R. 1994. Dilemmas in cooperative design. In *PDC'94: Proceedings of the Participatory Design Conference*, ed. R. Trigg et al. Computer Professionals for Social Responsibility.

Maxmen, J. S. 1987. Long term trends in health care: The post physician era reconsidered. In *Indicators and Trends in Health and Health Care*, ed. D. Schwefel. Springer.

May, W. E. 1985. Consensus or coercion. *Journal of the American Medical Association* 254: 1077.

Mayne, J. G., W. Weksel, and P. N. Sholtz. 1968. Toward automating the medical history. *Mayo Clinic Proceedings* 43: 1–25.

McDermott, W. 1971. Medicine in modern society. In *Cecil-Loeb Textbook of Medicine*, ed. P. Beeson and W. McDermott. Saunders.

McDonald, C. J. 1976. Protocol-based computer reminders, the quality of care and the non-perfectability of man. *New England Journal of Medicine* 295: 1351–1355.

McGehee Harvey, A., et al. 1984. *The Principles and Practice of Medicine*. Appleton.

McNeil, N., S. G. Pauker, H. C. Sox, and A. Tversky. 1982. On the elicitation of preferences for alternative therapies. *New England Journal of Medicine* 306: 1259–1262.

Meneely, G. R., O. Paul, H. F. Dorn, and T. R. Harrison. 1960. Cardiopulmonary semantics. *Journal of the American Medical Association* 174: 1628–1629.

Mesman, J. 1993. The digitalization of medical practice. Presented at Annual Meeting of the Social Studies of Science, West Lafayette, Indiana, 1994.

Miller, P. L. 1986. The evaluation of AI systems in medicine. *Computer Methods and Programs in Biomedicine* 22: 5–11.

Miller, R. A., and F. E. Masarie Jr. 1990. The demise of the "Greek oracle" model for medical diagnostic systems. *Methods of Information in Medicine* 29: 1–2.

Miller, R. A., M. A. McNeil, S. M. Challinor, F. E. Masarie, and J. D. Myers. 1986. The INTERNIST-1/QUICK MEDICAL REFERENCE project: Status report. *Western Journal of Medicine* 145: 816–822.

Miller, R. A., H. E. Pople, and J. D. Myers. 1982. INTERNIST-I, an experimental computer-based diagnostic consultant for general internal medicine. *New England Journal of Medicine* 307: 468–476.

Mirowski, P. 1992. When games grow deadly serious: The military influence on the evolution of game theory. In *Toward a History of Game Theory*, ed. E. Weintraub. Duke University Press.

Mol, A. 1993. Decisions no one decides about: Anemia in practice. Paper presented at Ethics in the Clinic (International Conference on Normative and Sociological Aspects of Clinical Decision Making), Maastricht.

Mol, A. Forthcoming. Missing links, making links: The performance of some atheroscleroses. In *Differences in Medicine*, ed. M. Berg and A. Mol.

Mol, A., and M. Berg. 1994. The principles and practices of medicine: The co-existence of various anemias. *Culture, Medicine and Psychiatry* 18: 247–265.

Mol, A., and J. Law. 1994. Regions, networks and fluids: Anaemia and social topology. *Social Studies of Science* 24: 641–671.

Morgan Jr., W. M. 1988. Clinical approach to the patient. In *Cecil Textbook of Medicine*, ed. J. Wijngaarden and L. Smith. Saunders.

Moser, R. H. 1974. An anti-intellectual movement in medicine? *Journal of the American Medical Association* 227: 432–434.

Moskowitz, A. J., B. J. Kuipers, and J. P. Kassirer. 1988. Dealing with uncertainty, risks, and tradeoffs in clinical decisions: A cognitive science approach. *Annals of Internal Medicine* 108: 435–449.

Mulkay, M. 1988. Don Quixote's double: A self-exemplifying text. In *Knowledge and Reflexivity*, ed. S. Woolgar. Sage.

Neal, M. P. 1951. Diagnostic drifts, deceptions and common misses. *Journal of the American Medical Association* 146: 537–541.

Neellon, F. A. 1974. POR relations. *New England Journal of Medicine* 290: 854–855.

Newell, A., and H. A. Simon. 1972. *Human Problem Solving*. Prentice-Hall.

Nyce, J. M., and W. Graves III. 1990. The construction of knowledge in neurology: Implications for hypermedia system development. *Artificial Intelligence in Medicine* 2: 315–322.

O'Connell, J. 1993. Metrology: The creation of universality by the circulation of particulars. *Social Studies of Science* 23: 129–173.

Oliver, M. F. 1985. Consensus or nonsensus conferences on coronary heart disease. *Lancet* May 11: 1087–1089.

Oughterson, A. W. 1955. Surgery, science and society. *New England Journal of Medicine* 252: 463–467.

Overall, J. E. 1972. Empirical approaches to classification. In *Computer Diagnosis and Diagnostic Methods*, ed. J. Jacquez. Thomas.

Pasveer, B. 1992. Shadows of Knowledge. Making a Representing Practice in Medicine: X-Ray Pictures and Pulmonary Tuberculosis, 1895–1930. Ph.D. thesis, University of Amsterdam.

Patil, R. S., P. Szolovits, and W. B. Schwartz. 1982. Modeling knowledge of the patient in acid-base and electrolyte disorders. In *Artificial Intelligence in Medicine*, ed. P. Szolovits. Westview.

Patterson, J. T. 1987. *The Dread Disease: Cancer and Modern American Culture*. Harvard University Press.

Pauker, S. G., G. A. Gorry, J. P. Kassirer, and W. B. Schwartz. 1976. Towards the simulation of clinical cognition: Taking a present illness by computer. *American Journal of Medicine* 60: 981–996.

Pearson, S. D., et al. 1992. The clinical algorithm nosology: A method for comparing algorithmic guidelines. *Medical Decision Making* 12: 123–131.

Pellegrino, E. D. 1964. The diagnostic process. I. In *The Diagnostic Process*, ed. J. Jacquez. Mallory.

Perrolle, J. A. 1991. Expert enhancement and replacement in computerized mental labor. *Science, Technology, and Human Values* 16: 195–207.

Pickering, A., ed. 1992. *Science as Practice and Culture*. University of Chicago Press.

Plante, D. A., J. P. Kassirer, D. A. Zarin, and S. G. Pauker. 1986. Clinical decision consultation service. *American Journal of Medicine* 80: 1169–1176.

Politser, P. 1981. Decision analysis and clinical judgment: A re-evaluation. *Medical Decision Making* 1: 361–389.

Pople, H. E. 1982. Heuristic methods for imposing structure on ill-structured problems: The structuring of medical diagnostics. In *Artificial Intelligence in Medicine*, ed. P. Szolovits. Westview.

Porter, T. M. 1995. *Trust in Numbers: The Pursuit of Objectivity in Science and Public Life*. Princeton University Press.

Potthoff, P., M. Rothemund, D. Schwefel, R. Engelbrecht, and W. van Eimeren. 1988. Expert systems in medicine: Possible future effects. *International Journal of Technology Assessment in Health Care* 4: 121–133.

Pylyshyn, Z. W. 1980. Computation and cognition: Issues in the foundations of cognitive science. *Behavioral and Brain Sciences* 3: 111–169.

Raiffa, H. 1968. *Decision Analysis: Introductory Lectures on Choices under Uncertainty*. Addison-Wesley.

Rees, C. 1981. Records and hospital routine. In *Medical Work*, ed. P. Atkinson and C. Heath. Gower.

Reggia, J. A., and S. Tuhrim, eds. 1985. *Computer-Assisted Medical Decision Making*. Springer-Verlag.

Reiser, S. J. 1978. *Medicine and the Reign of Technology*. Cambridge University Press.

Rennie, D. 1981. Consensus statements. *New England Journal of Medicine* 304: 665–666.

Reverby, S. 1981. Stealing the golden eggs: Ernest Amory Codman and the science and management of medicine. *Bulletin of the History of Medicine* 55: 156–171.

Richards, E. 1991. *Vitamin C and Cancer: Medicine or Politics?* St. Martin's.

Riesenberg, D. 1989. Economics is everybody's business. *Journal of the American Medical Association* 262: 2897.

Rizzo, J. A. 1993. Physician uncertainty and the art of persuasion. *Social Science and Medicine* 37: 1451–1459.

Robinson, M. 1991a. Computer supported co-operative work: Cases and concepts. Paper presented at *GROUPWARE '91*, Amsterdam.

Robinson, M. 1991b. Double-level languages and co-operative working. *AI and Society* 5: 34–60.

Robinson, M. 1994. Intimacy & abstraction & maps & terrains. Paper presented at conference on Locating Design, Development and Use, Oksnoen.

Rosenberg, C. E. 1987. *The Care of Strangers: The Rise of America's Hospital System*. Basic Books.

Rothman, D. J. 1991. *Strangers at the Bedside: A History of How Law and Bioethics Transformed Medical Decision Making*. Basic Books.

Ruby, A., and J. Morganroth. 1970. The need for a medical ideology. *Journal of the American Medical Association* 212: 2096–2097.

Rutstein, D. 1962. Another look at the patient. *New England Journal of Medicine* 267: 338–342.

Rutstein, D. D. 1961. Maintaining contact with medical knowledge. *New England Journal of Medicine* 265: 321–324.

Saetnan, A. R. 1991. Rigid politics and technological flexibility: The anatomy of a failed hospital innovation. *Science, Technology, and Human Values* 16: 419–447.

Savage, L. J. 1954. *The Foundations of Statistics*. Reprint: Dover, 1972.

Savage, L. J. 1972. Diagnosis and the Bayesian viewpoint. In *Computer Diagnosis and Diagnostic Methods*, ed. J. Jacquez. Thomas.

Schaffner, K. F., ed. 1985. *Logic of Discovery and Diagnosis in Medicine*. University of California Press.

Schoemaker, P. J. H. 1982. The expected utility model: Its variants, purposes, evidence and limitations. *Journal of Economic Literature* 20: 529–563.

Schwartz, W. B. 1970. Medicine and the computer: The promise and problems of change. *New England Journal of Medicine* 283: 1257–1264.

Schwartz, S. I. 1987. Tempering the technological diagnosis of appendicitis. *New England Journal of Medicine* 317: 703–704.

Schwartz, W. B. 1979. Decision analysis: A look at the chief complaints. *New England Journal of Medicine* 300: 556–559.

Schwartz, W. B., G. A. Gorry, J. P. Kassirer, and A. Essig. 1973. Decision analysis and clinical judgment. *American Journal of Medicine* 55: 459–472.

Schwartz, W. B., R. S. Patil, and P. Szolovits. 1987. Artificial intelligence in medicine: Where do we stand? *New England Journal of Medicine* 316: 685–688.

Searle, J. 1984. *Minds, Brains and Science.* Harvard University Press.

Shapin, S., and S. Schaffer. 1985. *Leviathan and the Air-Pump: Hobbes, Boyle, and the Experimental Life.* Princeton University Press.

Shortliffe, E. H. 1987. Computer programs to support clinical decision making. *Journal of the American Medical Association* 258: 61–66.

Shortliffe, E. H. 1991. Medical informatics and clinical decision making: The science and the pragmatics. *Medical Decision Making* 11 (Suppl.): 2–14.

Shortliffe, E. H., S. G. Axline, B. G. Buchanan, T. C. Merigan, and S. N. Cohen. 1973. An artificial intelligence program to advise physicians regarding antimicrobial therapy. *Computers and Biomedical Research* 6: 544–560.

Shortliffe, E. H., B. G. Buchanan, and E. A. Feigenbaum. 1979. Knowledge engineering for medical decision making: A review of computer-based clinical decision aids. *Proceedings of the IEEE* 67: 1207–1224.

Shortliffe, E. H. et al. 1984. An expert system for oncology protocol management. In *Rule-Based Expert Systems,* ed. B. Buchanan and E. Shortliffe. Addison-Wesley.

Silverman, D. 1987. *Communication in Medical Practice.* Sage.

Silverstein, M. D. 1988. Prediction instruments and clinical judgment in critical care. *Journal of the American Medical Association* 260: 1758–1759.

Simon, H. A. 1979. From substantive to procedural rationality. In *Philosophy and Economic Theory,* ed. F. Hahn and M. Hollis. Oxford University Press.

Simon, H. A. 1981. *The Sciences of the Artificial.* MIT Press.

Simon, H. A. 1986. Alternative visions of rationality. In *Judgment and Decision Making,* ed. H. Arkes and K. Hammond. Cambridge University Press.

Singleton, V. Forthcoming. Stabilizing instabilities: The laboratory in the UK cervical screening program. In *Differences in Medicine,* ed. M. Berg and A. Mol.

Slack, W. V. 1972. Patient power: A patient-oriented value system. In *Computer Diagnosis and Diagnostic Methods,* ed. J. Jacquez. Thomas.

Smith, L. H. 1988. Medicine as an art. In *Cecil Textbook of Medicine,* ed. J. Wyngaarden and L. Smith. Saunders.

Smith, D. 1990. *Texts, Facts, and Femininity: Exploring the Relations of Ruling.* Routledge.

Society for Medical Decision Making Committee on Standardization of Clinical Algorithms. 1992. Proposal for clinical algorithm standards. *Medical Decision Making* 12: 149–154.

Sokal, R. R. 1964. Numerical taxonomy and disease classification. In *The Diagnostic Process,* ed. J. Jacquez. Mallory.

Southgate, M. T. 1975. Demythologizing medicine; or, how to stop worrying and love the computer. *Journal of the American Medical Association* 232: 515–516.

Sox, H. C. Jr., C. H. Sox, and R. K. Tompkins. 1973. The training of physician's assistants: The use of a clinical algorithm system for patient care, audit of performance and education. *New England Journal of Medicine* 288: 818–824.

Star, S. L. 1989a. Layered space, formal representations and long-distance control: The politics of information. *Fundamenta Scientiae* 10: 125–154.

Star, S. L. 1989b. The structure of ill-structured solutions: Boundary objects and heterogeneous distributed problem solving. In *Distributed Artificial Intelligence Vol. 2*, ed. M. Huhns and L. Gasser. Pitman.

Star, S. L. 1991. Power, technologies and the phenomenology of conventions: On being allergic to onions. In *A Sociology of Monsters*, ed. J. Law. Routledge.

Star, S. L. 1992. Craft vs. commodity, mess vs. transcendence: How the right tool became the wrong one in the case of taxidermy and natural history. In *The Right Tools for the Job*, ed. A. Clark and J. Fujimura. Princeton University Press.

Star, S. L., ed. 1995. *Ecologies of Knowledge: Work and Politics in Science and Technology*. State University of New York Press.

Star, S. L., and J. R. Griesemer. 1989. Institutional ecology, "translations," and boundary objects: Amateurs and professionals in Berkely's Museum of Vertebrate Zoology, 1907–39. *Social Studies of Science* 19: 387–420.

Starr, P. 1982. *The Social Transformation of American Medicine*. Basic Books.

Stevens, R. 1989. *In Sickness and in Wealth: American Hospitals in the Twentieth Century*. Basic Books.

Stoop, A., and M. Berg. 1994. Spiegels voor de arts. Beslissingsondersteunende technieken en medische praktijken. *Gezondheid: Theorie in Praktijk* 2: 265–278.

Strasser, P. H., J. C. Levy, G. A. Lamb, and J. Rosekrans. 1979. Controlled clinical trial of pediatric telephone protocols. *Pediatrics* 64: 553–557.

Strauss, A., S. Fagerhaugh, B. Suczek, and C. Wieder. 1985. *Social Organization of Medical Work*. University of Chicago Press.

Suchman, L. 1987. *Plans and Situated Actions: The Problem of Human-Machine Communication*. Cambridge University Press.

Suchman, L. 1993a. Centers of coordination: A case and some themes. Paper presented at *NATO Advanced Research Workshop on Discourse, Tools and Reasoning*, Lucca, Italy.

Suchman, L. 1993b. Technologies of accountability: Of lizards and aeroplanes. In *Technology in Working Order*, ed. G. Button. Routledge.

Suchman, L. 1994. Working relations of technology production and use. *Computer Supported Cooperative Work* 2: 21–40.

Suchman, L., and R. H. Trigg. 1993. Artificial intelligence as craftwork. In *Understanding Practice*, ed. S. Chaiklin and J. Lave. Cambridge University Press.

Sultan, C., M. Imbert, and G. Priolet. 1988. Decision-making system (DMS) applied to hematology: Diagnosis of 180 cases of anemia secondary to a variety of hematologic disorders. *Hematologic Pathology* 2: 221–228.

Sutherland, J. W. 1986. Assessing the artificial intelligence contribution to decision technology. *IEEE Transactions on Systems, Man and Cybernetics* 16: 3–20.

Szolovits, P., ed. 1982. *Artificial Intelligence and Medicine*. Westview.

Szolovits, P., and W. J. Long. 1982. The development of clinical expertise in the computer. In *Artificial Intelligence and Medicine*, ed. P. Szolovits. Westview.

Szolovits, P., R. S. Patil, and W. B. Schwartz. 1988. Artificial intelligence in medical diagnosis. *Annals of Internal Medicine* 108: 80–87.

Szolovits, P., and S. G. Pauker. 1978. Categorical and probabilistic reasoning in medical diagnosis. *Artificial Intelligence* 11: 115–144.

Talbott, J. H. 1961. The art of medicine. *Journal of the American Medical Association* 175: 898–899.

Talmon, J. L., R. A. J. Schijven, P. J. E. H. M. Kitslaar, and R. Penders. 1987. An expert system for diagnosis and therapy planning in patients with peripheral vascular disease. In *Proceedings of European Conference on Artificial Intelligence in Medicine*, ed. J. Fox et al. Springer-Verlag.

Tielens, V. 1989. Standaarden: het gezicht van de huisarts. *Huisarts & Wetenschap* 32: 3.

Timmermans, S. 1993. Calling the Patient: Interpretative Information Work and the Closure of a Resuscitative Trajectory. Unpublished manuscript, University of Illinois, Champaign.

Timmermans, S., and M. Berg. 1996. Standardization in action: Achieving universalism and localization in medical protocols. *Social Studies of Science*, in press.

Timmermans, S., G. C. Bowker, and S. L. Star. Forthcoming. The architecture of difference: visibility, controllability, and comparability in building a nursing intervention classification. In *Differences in Medicine*, ed. M. Berg and A. Mol.

Tompkins, R. K., W. D. Kniffin Jr., H. C. Sox Jr., C. H. Sox, and A. D. Kaplan. 1973. Use of a clinical algorithm system in a physician's assistant program. In *Applying the Problem-Oriented System*, ed. H. Walker et al. Medcom.

Turkle, S. 1984. *The Second Self: The Human Spirit in a Computer Culture*. Simon & Schuster.

Tversky, A., and D. Kahneman. 1974. Judgment under uncertainty: Heuristics and biases. *Science* 185: 1124–1131.

Vaisrub, S. 1971. Etched in twilight. *Journal of the American Medical Association* 215: 1318.

van Bemmel, J. H., and J. L. Willems, eds. 1989. *Handboek Medische Informatica.* Bohn, Scheltema en Holkema.

van Dijk, P. H., and J. Bomhof. 1991. Consensusontwikkeling: Een basisbehoefte? *Medisch Contact* 46: 345–348.

Van Way III, C. W., J. R. Murphy, E. L. Dunn, and S. C. Elerding. 1982. A feasibility study of computer aided diagnosis in appendicitis. *Surgery, Gynecology and Obstetrics* 155: 685–688.

Vandenbroucke, J. P. 1988. Kwantitatieve nieuwlichterij in de geneeskunde: Een poging tot ordening. *Nederlands Tijdschrift voor de Geneeskunde* 132: 337–340.

Vang, J. 1988. The consensus development conference and the European experience. *International Journal of Technology Assessment in Health Care* 4: 65–76.

Vogel, M. J., and C. E. Rosenberg, eds. 1979. *The Therapeutic Revolution: Essays in the Social History of American Medicine.* University of Pennsylvania Press.

Wachtel, T., A. W. Moulton, J. Pezzullo, and M. Hamolsky. 1986. Inpatient management protocols to reduce health care costs. *Medical Decision Making* 6: 101–109.

Wagner, I. 1994. Hard times: The politics of women's work in computerized environments. Paper presented at Fifth IFIP Working Conference on Women, Work and Computerization, Manchester.

Warndorff, D. K., C. J. M. Pouls, and J. A. Knottnerus. 1988. Medische besliskunde in Nederland: Een literatuuronderzoek. *Medisch Contact* 43: 1174–1178.

Warner, H. R., A. F. Toronto, and L. G. Veasy. 1964. Experience with Bayes' theorem for computer diagnosis of congenital heart disease. *Annals of the New York Academia of Science* 115: 558–567.

Warner, J. H. 1985. Science in medicine. *Osiris* (second series) 1: 37–58.

Warner, J. H. 1986. *The Therapeutic Perspective: Medical Practice, Knowledge and Identity in America, 1820–1885.* Harvard University Press.

Weaver, R. R. 1991. *Computers and Medical Knowledge: The Diffusion of Decision Support Technology.* Westview.

Webster, J. 1991. Advanced manufacturing technologies: Work organisation and social relations crystallised. In *A Sociology of Monsters*, ed. J. Law. Routledge.

Weed, L. L. 1968. Medical records that guide and teach. *New England Journal of Medicine* 278: 593–600, 652–657.

Weed, L. L. 1971. *Medical Records, Medical Education, and Patient Care.* Year Book Medical Publishers.

Weinstein, M. C. et al. 1980. *Clinical Decision Analysis.* Saunders.

Weizenbaum, J. 1976. *Computer Power and Human Reason: From Judgment to Calculation.* Freeman.

Wennberg, J. 1984. Dealing with medical practice variations: A proposal for action. *Health Affairs* 3: 6–32.

Wennberg, J. E. 1991. Unwanted variations in the rules of practice. *Journal of the American Medical Association* 265: 1306–1307.

Whalen, J. 1993. Accounting for "standard" task performance in the execution of 9-1-1 operations. Paper presented at the *Annual Meetings of the American Sociological Association*, Miami, August 1993.

White, I. 1988. W(h)ither expert systems? A View from the Outside. *AI and Society* 2: 161–171.

Wiersma, T. 1992. Overwegingen bij de NHG-standaard cholesterol. *Huisarts en Wetenschap* 35: 97–100.

Willems, D. 1995. Tools of Care: Explorations into the Semiotics of Medical Technology. Ph.D. thesis, University of Limburg.

Williamson, J. W. 1973. Evaluating the quality of medical care. *New England Journal of Medicine* 288: 1352–1353.

Wilson, D. H., P. D. Wilson, R. G. Walmsley, J. C. Horrocks, and F. T. de Dombal. 1977. Diagnosis of acute abdominal pain in the accident and emergency department. *British Journal of Surgery* 64: 250–254.

Winickoff, R. N., K. McCue, and F. Perlman. 1977. Upper respiratory infection. In *A Problem-Oriented Textbook with Protocols*, ed. A. Komaroff and R. Winickoff. Little, Brown.

Winograd, T., and F. Flores. 1986. *Understanding Computers and Cognition: A New Foundation for Design*. Ablex.

Wittgenstein, L. 1958. *Philosophical Investigations*. Blackwell.

Wood, D. 1992. *The Power of Maps*. Guilford.

Wooffit, R., and N. Fraser. 1993. We're off to ring the wizard, the wonderful Wizard of Oz. In *Technology in Working Order*, ed. G. Button. Routledge.

Woolgar, S. 1985. Why not a sociology of machines? The case of sociology and artificial intelligence. *Sociology* 19: 557–572.

Woolgar, S. 1987. Reconstructing man and machine: a note on sociological critiques of cognitivism. In *The Social Construction of Technological Systems*, ed. W. Bijker et al. MIT Press.

Woolgar, S. 1988. *Science: The Very Idea*. Ellis Horwood.

Woolgar, S. 1989. Representation, cognition and self: What hope for an integration of psychology and sociology? In *The Cognitive Turn*, ed. S. Fuller. Kluwer.

Woolgar, S. 1991. Configuring the user: The case of usability trials. In *A Sociology of Monsters*, ed. J. Law. Routledge.

Wouters, P. 1992. Een nieuw teken aan de wand: De uitvinding van de Science Citation Index 1948–1964. *Kennis en Methode* 16: 313–336.

Wulff, H. R. 1981. *Rational Diagnosis and Treatment.* Blackwell.

Wyatt, J. 1989. Lessons learned from the field trial of ACORN, an expert system to advise on chest pain. In *Proceedings Medical Informatics 1989, Singapore,* ed. P. Manning et al. Elsevier.

Wyatt, J. 1991a. *A Method for Developing Medical Decision Aids Applied to ACORN, a Chest Pain Advisor.* University of Oxford.

Wyatt, J. 1991b. Use and sources of medical knowledge. *Lancet* 338, November 30: 1368–1373.

Wyatt, J., and P. Emerson. 1990. A pragmatic approach to knowledge engineering with examples of use in a difficult domain. In *Expert Systems,* ed. D. Berry and A. Hart. MIT Press.

Wyatt, J., and D. Spiegelhalter. 1991. Evaluating medical expert systems: What to test, and how? In *Knowledge Based Systems in Medicine* (proceedings of workshop on System Engineering in Medicine, Maastricht, 1989).

Wyngaarden, J. B. 1979. Biological science and medical practice. In *Cecil Textbook of Medicine,* ed. P. Beeson et al. Saunders.

Wynne, A. 1988. Accounting for accounts of the diagnosis of multiple sclerosis. In *Knowledge and Reflexivity,* ed. S. Woolgar. Sage.

Wynne, B. 1988. Unruly technology: Practical rules, impractical discourses and public understanding. *Social Studies of Science* 18: 147–167.

Young, A. 1981a. The creation of medical knowledge: Some problems in interpretation. *Social Science and Medicine* 15B: 379–386.

Young, A. 1981b. When rational men fall sick: An inquiry into some assumptions made by medical anthropologists. *Culture, Medicine and Psychiatry* 5: 317–335.

Yu, V. L. et al. 1979. Antimicrobial selection by a computer. *Journal of the American Medical Association* 242: 1279–1282.

Zerubavel, E. 1979. *Patterns of Time in Hospital Life.* University of Chicago Press.

Zimmerman, D. H. 1970. The practicalities of rule use. In *Understanding Everyday Life,* ed. J. Douglas. Aldine.

Zuboff, S. 1988. *In the Age of the Smart Machine: The Future of Work and Power.* Basic Books.

Zubrod, C. 1984. Origins and development of chemotherapy research at the National Cancer Institute. *Cancer Treatment Reports* 68: 9–19.

Zwaard, A., S. Zijstra, and R. Grol. 1989. Kwaliteits- en deskundigheidsbevordering rond NHG standaarden. *Huisarts en Wetenschap* 32: 501–504.

Index

Action
  behavior-specific, 202
  regular, 202
Actor-network theory, 189–190, 202
Agency for Health Care Policy and
    Research, 187
Algorithms. *See* Protocols
Anspach, Renée, 100–102
Artificial intelligence, 4–5, 65–66, 74,
    113, 155, 180, 188

Bayes' Theorem, 2, 43, 46, 54–56,
    81–84, 112–113, 141, 155, 184–185,
    189
  independence assumption of, 54, 70,
    187–188
  mutual exclusivity assumption of, 54,
    188
Bleich, Howard, 107

Callon, Michel, 189–190
Centraal Begeleidingsorgaan voor de
    Intercollegiale Toetsing (CBO),
    188
Ceteris paribus condition, 5–6, 160
Collins, Harry, 162, 201–202
Computer
  as metaphor for mind, 39–41, 45, 65,
    74, 77, 163
  as symbol of science, 66
Computer-based decision-support
    techniques, 3–7, 33, 91, 97, 105–108,
    110, 116–117, 120, 150, 155–163, 171,
    175, 199. *See also* Decision-support
    techniques; Expert systems;
    Statistical tools
ACORN, 80–102, 105–121, 142–147,
    150, 153, 155, 171–173, 190
acute abdominal pain tool, 1–2, 5–6,
    41–46, 49–50, 81, 89–97, 101,
    105–113, 139–142, 147–150, 155,
    160, 167, 171, 186–187, 192, 203–204
GLADYS, 141–142, 148, 198
INTERNIST, 155, 188, 194
MYCIN, 66–70, 74–75, 155, 158–159,
    189, 191, 194, 199, 201
ONCOCIN, 194, 201
peripheral vascular disease system,
    198–199
PIP, 188
Problem Knowledge Computer, 182
Consensus reports. *See* Protocols
Convergence, 118, 165–168, 176
Cooperative work, computer-supported,
    179, 204

Decision analysis, 3–7, 33–34, 46–52,
    63–64, 74–76, 103–104, 114,
    155–161, 187–192, 196, 200
  limits of, 6–7, 50–51, 54–57, 70–71
Decision making. *See* Medical decision
    making
Decision-support techniques. *See also*
    Computer-based decision-support
    techniques; Decision analysis; Diag-
    nosis; Formal systems; Protocols;
    Statistical tools
  evaluation of, 171–172

Decision-support techniques (cont.)
  politics of, 119–121, 173–177
  success of, 155–165, 169–170, 176, 190, 201
de Dombal, F. T., 1–2, 41–46, 49–50, 94, 96, 159, 167, 193
Design as critique, 174–178
Diagnosis, statistical, 29–30, 41–52, 184. *See also* Medical decision making
Disciplining of practice to formalism, 89–102, 104, 109–121, 123–124, 140–141, 144, 147, 163–170, 176–178, 202
Dreyfus, Hubert, 5–7, 79, 161–162, 202

Economics, and medicine, 29, 46
Eddy, David, 2, 32–34, 52, 63–64, 103–104, 109–114, 188
Elstein, Arthur, 27
Emerson, Peter, 81–86, 107, 112, 150, 155, 190
Ethnomethodology, 131
Expert systems, 5, 7, 65–77, 79, 83, 105, 112–113, 116–119, 139–140, 155, 162–164, 177, 188–189

Feinstein, Alvan, 22–25, 30, 34–36, 53–57, 60–62, 70–71, 76–77, 168, 176, 184, 186–189, 193
Formal domains, 5, 80, 91–92, 113, 202
Formal systems, 5, 9, 11, 80, 88–102, 113–114, 118–120, 123–124, 138–153, 161–169, 177–178, 202. *See also* Decision-support techniques
  accumulating, 139–140, 148–153, 171, 177, 202
  coordinating, 138–140, 148–153, 171, 177, 202
  disciplining of practice to, 89–102, 104, 109–121, 123–124, 140–141, 144, 147, 163–170, 176–178, 202
  limits of, 5–7, 160–165, 170–172, 175
  as maps, 162–178

Greenfield, Sally, 57–61

Guidelines. *See* Protocols

Human-computer interface, 159
Hutchins, Edwin, 133

Information crisis in American science, 181

Judgment, clinical. *See* Medical decision making; Medical practice

Komaroff, Anthony, 11, 34–36

Latour, Bruno, 189–190
Ledley, Robert, 41
Localization of a tool, 104–121, 140, 151–153, 165–170, 175, 178, 195, 197
  in rationale, 104, 111–115, 119
  in scope, 104, 110–111, 114–115, 195
  in space, 104–110, 114–115
  in time, 194
Logics, 117, 150–151, 167–168, 172–178
  action-sequence-focused, 61–63, 76–77
  clinical, 54–77, 112–113, 116–117, 170
  decision-focused, 61, 63, 71–77, 117–119, 173–174
  statistical, 45–77, 112–113, 116–117, 167, 170, 187
Lusted, Lee, 2, 41

Management, scientific, 26, 186
Materialization of a tool's exigencies, 93–95, 97–98, 117
Medical criteria
  changes in, 98–101, 115–118, 153, 171–174, 176–177
  (re)construction of, 130–138, 144–151, 197–199
  vagueness of, 55
Medical data
  (re)construction of, 127–138, 141–144, 147–151, 196
  soft vs. hard, 55–56, 61, 100–102, 176
  subjective vs. objective, 100–102

Medical decision making
  control of, 7–8, 124, 132–140,
    148–153, 161, 170–173, 177
  nature of, 4–7, 27–37, 44–47, 50–57,
    63–65, 70–77, 79, 135–138,
    161–174, 177, 183, 201
  problems of, 2–5, 11, 15–16, 21,
    31–37, 45, 50–53, 75–77, 80–84,
    173–174, 183–184
Medical Decision Making Society, 174
Medical education, 13, 26–27, 155, 183
Medical personnel. *See also* Physicians
  influence of, 95–98, 101–102, 120–121
    (*see also* Medical decision making;
    Medical practice)
  reactive, opportunistic stand of, 135
  responsibilities of, 86, 93, 101,
    120–121, 124, 148–151, 172, 177
  skill of (*see* Medical practice)
  and a tool's demands, 140–153,
    166–169
Medical practice
  anticipation and, 146–147, 151
  as art, 4–5, 8, 11–15, 18, 25, 31, 34,
    36, 180–182
  control of by outsiders, 7, 16, 26, 35,
    63, 65, 118–119
  costs of, 11, 15–18, 26, 29, 34, 36, 46,
    74, 103–104, 120–121, 157, 176
  dehumanization of, 7, 14–15, 55,
    100–102, 120–121, 175
  hierarchy in, 92–94, 97–98, 117,
    120–121
  nature of, 8, 11–37, 39–41, 44–47,
    50–66, 70–77, 79, 123–147,
    161–174, 201
  problems of, 2–5, 8, 11–37, 39–41,
    44–45, 50–53, 75–77, 80–81, 151
  quality of, 2–5, 11, 15–19, 26, 32, 34,
    36, 64
  rationality of (*see* Logics)
  as science, 4–5, 8, 11–37, 39, 45–46,
    53–57, 60–61, 75–77, 79, 119, 162,
    173, 180–183
  and skills, 4–9, 14–15, 20, 36, 63, 90,
    105, 120–121, 160, 162, 170, 177, 201
  socioeconomic context of, 15–19,
    34, 201
  as terrain, 162–178
  and uncertainty, 4–5, 32, 35, 46–52,
    63, 71
  variations in, 2–5, 21, 32, 63, 88, 96,
    116–117, 139, 173
Medical profession, 11–19, 26–27, 29,
    40, 64, 180–184, 201
Medical record
  electronic, 9, 23, 177, 181
  problem-oriented, 23, 33, 182
  standardization of, 20–26, 33, 35
Medical research, 13–15, 19–22, 35, 49,
    53, 64, 80, 87–88, 97, 103, 139, 156,
    164, 181–182, 192, 194, 200
  and statistics, 40, 46, 64, 76, 184
Medical terminology
  chaos of, 20, 24, 35, 55, 159, 163,
    184, 193
  standardization of, 21–26, 33

National Health Service, 181
Nederlands Huisartsen Genootschap,
    188
Newell, Alan, 27, 65–66

Patient(s)
  influence of, 95–96, 101, 120–121
  perspective of, 49, 189, 204
  trajectory of, 124–140, 149–153, 166,
    169, 176, 196
Pauker, Stephen, 189, 200
Physicians. *See also* Medical decision
    making; Medical personnel;
    Medical practice; Medical profession
  as central actors, 132–134, 137–138
  as information processors, 27–37,
    65, 74, 162–163, 184
  as "men of character," 13–19, 31, 34
  number of, 13, 15
  as statistical reasoners, 29–37, 45–47,
    50–52, 162–163, 184
Plateau and cliff effect, 110, 119, 199
Practice policies. *See* Protocols
Probability, 4, 29–30, 32–35, 41–52, 54–
    57, 63, 70, 75–76, 83, 106, 183–186
  objective vs. subjective, 49–51, 76

238   Index

Promise of science, belief in, 13, 26, 40, 181–182
Protocols, 3–7, 25–26, 34–35, 52–65, 76–77, 79, 92–97, 105–120, 131, 138–140, 144–150, 153, 156–162, 172–177, 199. *See also* Decision-support techniques
  for breast cancer, 80–81, 86–102, 108–121, 174
  consensus reports, 63–65, 75–76, 106, 108, 112, 114, 156–160, 194
  for contrast agents, 103–105, 109–116
  for evaluation of disability, 19
  limits of, 7, 63–65, 71, 74
  for lung cancer, 143, 146–149
  of National Institute of Health, 64, 156, 159
  for physician extenders, 57–61, 65, 94–95
  for research, 53, 80, 94, 97, 101, 105–106, 109, 120, 139, 150–151, 156, 161, 164, 200, 204
  for upper-respiratory-tract complaints, 57–61, 93, 105, 112, 194–195
Psychology
  as a science, 39–41
  cognitive, 27–37, 39–41, 50, 65, 76, 164, 184–185
  statistical methods in, 39–41

Raiffa, Howard, 46
Rules, conditional, 3–6, 67, 79, 83–84, 88–91, 117–119, 161, 175, 186, 191

Savage, Leonard, 46
Schwartz, William, 155
Science and technology studies, 8, 189–190
Script of technology, 86–88, 105–108, 113–114, 118, 121, 140–142
Sensitivity analysis, 186
Shortliffe, Edward, 66–71, 158–159
Simon, Herbert, 27, 65–66
Socialized medicine, 15–16, 181
Spokesmanship, 86–87, 190
  reshuffling of, 95–97, 117–121

Standardization, 20–27, 33–34, 53–56, 60–61, 74, 76–77, 80, 90, 97–98, 114, 118, 141, 161, 166, 169, 177, 181. *See also specific topics*
Standards. *See* Protocols; Standardization
Statistical tools, 39–57, 61, 71, 76–77, 79–84, 100, 105, 108, 116–119, 139–140, 155, 162–163, 171, 188. *See also* Computer-based decision-support techniques; Decision analysis; Decision-support techniques
  as metaphor for mind, 39–41, 45–46, 63–64, 77
  limits of, 46–50, 53–57, 70–71
Suchman, Lucy, 176

Taylor, Frederick, 26, 186
Trustworthiness, 95–96, 141

Universality, 7, 39–40, 63–64, 74, 104, 114, 121, 152, 158–163, 167–170, 175, 194–195
Utility, 4, 29, 46–52, 54–55, 63, 70, 75–76, 158, 183–184, 186
  measurement of, 51, 103–104, 186

Weed, Lawrence, 22–25, 30, 34, 35, 53, 77, 182, 186
Wittgenstein, Ludwig, 6
Wyatt, Jeremy, 80–86, 150, 190